21st CENTURY JJEDI

Spiritual Athletes on the Cosmic Chessboard

Next Generation Edition

By

Dr. Damon Sprock

TM

San Diego, CA USA

Copyright: 2018, 3rd Edition, by Dr. Damon Sprock

Library of Congress   Washington, DC

Sprock, Damon: 21ST CENTURY JJEDI: Spiritual Athletes on the Cosmic Chessboard, Next Generation Edition

Published by: Jupiter Productions – San Diego, CA,   USA

TABLE OF CONTENTS

AUTHOR BIO

The year was 1971, two years after I graduated from Purdue University. My journey into the metaphysical was about to begin in a most unusual but blessed manner. My best friend and I were enjoying a Sunday afternoon in San Francisco when two pretty girls asked us if we wanted to join them at their houseboat in Sausalito, just across the Golden Gate Bridge. We agreed, and off we went. When we arrived at their houseboat, I was taken aback by its enormity, 123 feet. It turned out that the houseboat was the Vallejo, a former tour, ferry boat built in Portland, Oregon in 1879. When I asked the girl if it was their boat, she said they were staying there with Alan Watts and his nephew, an artist. I looked at her and asked, "Are you speaking about Alan Watts, the internationally, renowned philosopher from England?" She answered, "Yes, you've heard of him?" I exclaimed, "Yes! I have been following his lectures." What a day; I met my future mentor and discovered my purpose.

My journey took me through many years of research into the depths of universal structure, many times losing myself in detailed visualization and imagination. It was difficult discussing with others the evidence I was unveiling regarding how the universe was structured, especially the quantum aspects. I learned to keep my research to myself. I prayed to God for guidance and protect me while I proceeded from anointment to appointment, knowing that between the promise and the fulfillment is grace.

My quest was to discover who God is, His location in the universe, how the universe was created by a spiritual entity, and how the consciousness of humanity has its origin from God's infinite, irreducible, spiritual dimension, a quantum connection that occurs through universal, coded frequency and inserts into our subconscious mind hologram. This enables the subconscious mind hologram to contain all preexisting knowledge, data that can be accessed and applied to our brain for greater decision-making and problem-solving. Understanding the function of this universal system can produce a huge up-grade in the hardware of our brain and the software of the subconscious mind hologram.

The discovery of this knowledge takes the "Law of Attraction" to the highest degree, that being how the law functions at the quantum level. This revelation allows us to now have access to all pre-existing, potential knowledge and gives humanity the boost needed to surge beyond the 10-12% brain usage at this point in our history.

I began my research, to send a message to the world that, we are at a position of critical mass with regard to our human condition and the outcome of the planet will be determined by our behavior. I have been at the perimeter of human consciousness since 1982. The research I have performed and applied is continuously setting new standards in Science of Mind and Spirituality. I hope you find my discoveries enlightening and the systems applicable to your path in life.

My interests in the mind began at age ten when an application (Visualizations influencing the subconscious mind hologram) of the mind sciences enabled me to achieve a record fifteen regular-season home runs in Little League baseball. From then on, it was one accomplishment after another. I went on to earn a football scholarship to Purdue University, graduating with a B.S. in Physical Therapy. Advanced studies in the field of Metaphysics/Cosmology/Ontology and Eastern philosophy highlighted my studies in California from 1980-1983.

My 1984 best-selling book, THE ATHLETIC CONNECTION and the techniques contained within have influenced, in some manner, every Olympic athlete since 1984 (A copy of the book was sent to every governing body of the U.S.

Olympic team.), along with such notables as: Jim Brown (NFL Hall of Fame), Oprah Winfrey, my cousin, Walt Corey (Super Bowl IV Champion and Former defensive coordinator for the Buffalo Bills), Gene Tenace (Former Oakland A's triple World Series winner and Toronto Blue Jays batting coach during and after their 1992 and 1993 World Series championships), to name a few.

FORWARD

When Dr. Damon Sprock was my special guest presenter at a meeting of the North San Diego County Institute of Noetic Science, cosponsored by the Creative Center for Spiritual Living of Escondido, we could not have possibly known what an empowering gift he had in store for us. Evidence of Intentionality and Non-Randomness...Dr. Damon Sprock facilitates his Bi-Mass/Velocity Factor and UNIFICATION PROCLAMATION: HUMAN CONSCIOUSNESS. Dr. Sprock introduces his "Unification Proclamation: Human Consciousness" ...... evidence of interconnectedness between all systems of knowledge, intention, and manifestation regarding human consciousness. Once this connection is known and embraced, one is able to leap into a more cooperative stage of relationships and attract unlimited levels of abundance. In this life-guide, you will be introduced to the location, function, and insertion of universal consciousness as it applies to your own unique, human awareness.

With an increased appreciation for why synchronous and often repetitive circumstances, encounters, and related behaviors occur, beneficiaries are empowered to choose anew, rather than continually recall and repeat undesired patterns of resistance-attraction and self-sabotage. Inclusive with and regardless of a person's religious or spiritual beliefs, ones who practice and apply the avant-garde principles diagrammed by Dr. Sprock become beneficiaries of the foretold *Heaven on Earth* that is already within each of us...awaiting only acceptance-by-degree.

Dr. Larry James Stevens: Director of Education for the Dodona Human Potential Research Institute, Community Coordinator for the Institute of Noetic Science (www.noetic.org) Author of "Celestial Fire – A Naval Aviator's Spiritual Odyssey" www.soulmanlarry.com

MIRACLES IN THE MAKING

In the 1960s, America was grasped by a new consciousness that would change its view of the very basic concepts for which it was established. In the effort to create for America a new alternative for the betterment of all citizens, the new "train of freedom" was side-tracked by a series of cause and effect circumstances that has brought us to a critical dilemma. Finite desires for lifestyles based on instant gratification were imposed upon the masses by a few whose underlying motives are to gain control of and maintain the hypnotic manipulation that follows social and economic variance.

It is difficult to resist the everyday temptations of physical world desire, especially when we are continuously exposed to the pleasures they bring. The 1960s revealed the freedom of these worldly desires to such an extreme that, an imbalance in our quest to achieve them occurred that has brought America to its present state, a consciousness of mental and emotional chaos. When an imbalance of this magnitude imposes itself and is accepted by masses of individuals on a daily basis, a chain of events begins that result in the destruction of the very core of humankind - the soul.

We are at the crossroads in the history of humankind at this moment. There is a choice to be made, and the responsibility belongs to us all. Each of us, in our own heart, has to come to the realization that, in order for our species to continue to exist, we must reverse the mental/emotional status of our consciousness and return equilibrium to our actions.

In the chapters that follow, it is my ultimate goal to reveal to you a philosophy for the new millennium, a philosophy that computes the answers to the daily strife we all experience and that when applied on a daily basis, can unveil the true freedom beneath the façade of physical world phenomenon.

"And the New York Times said, 'God is dead.'"[1]. God is dead, a sad revelation for a society now five decades entranced into the 1960s post-enlightenment era. God is dead, the battle cries of the new atheism by the ego-entrenched hordes that have become entwined in a mixture of flesh-controlled mentality and weak emotion.

Fool-hearted ones, God is not dead. God is allowing you to experience your lives of over-kill, emotional reaction. Without this knowledge logged into your consciousness, you will not be able to take the next step in your evolution of consciousness. This is God's grand plan. When God is ready to have you realize this truth, and this will occur according to God's infinite timetable, then will humankind begin reacting to Nature not with high degrees of emotional content but with spiritual intent.

To further explain this façade that we all experience in life, I will give you an analogy of how the facade works. A movie was made recently entitled, THE MATRIX. In the movie, a computer-generated world was created by machines for the sole purpose of deceiving humankind into believing the world they were experiencing was reality. Our lives are also being deceived, and the deceiver is our own <u>perception</u> of Nature.

We perceive Nature to be the matrix, the grand illusion of our five senses; however, Nature is God's perfect design. It is <u>we</u> who have to adjust to <u>it</u>. Nature teases us with its bits and pieces of fulfillment, only to create circumstances that allow our desires to slip through our clutches. Nature seems like a magician whose acts of appearing and disappearing keeps us ever-chasing the dangling "carrot of desire." We are like the fly that flounders against the window, trying desperately to get back to freedom, only to become crazed with desire and finally die. We like the fly need to seek another path back to reality. We must move away from the transparency, our perception of reality, and release our stronghold of illusions we have about Nature. To do this is to give ourselves true freedom within.

Dr. Damon Sprock

PREFACE

We are entering a new Grand Cycle of transformation in the evolution of human consciousness. This Grand Cycle occurs every twenty-four thousand years with intermediate cycles every two thousand years, the last, great breakthrough in the human behavior model being revealed by Jesus Christ. It has carried us to this point. However, along the way, important values have been lost or deviated from in translations and actions. We need a new paradigm, something pragmatic in our modern day society. Without a clear understanding of how life works and why we are really here, the elements of Nature will continue to appear to be expounding its matrix-like influence upon us, keeping humankind in a state of flux and chaos.

In "21ST CENTURY JJEDI," a new generation of readers are informed of their true potentials, how to perceive reality based on cutting-edge, scientific evidence, ancient wisdom, and how to react to those realities in order to raise the quality of their consciousness and thus their lifestyles. One of the first laws, The Law of Attraction, is only one law that is needed for your ascension. It is essential to understand and apply other laws in order for the Law of Attraction to take full effect. Knowledge of the underlying factors of universal structure is revealed to you. This is the first time in history that the quantum pathway connecting God consciousness to human consciousness via our subconscious mind hologram is revealed. The dots have been connected.

With this new paradigm, humanity can now complete the maturation stage of consciousness evolution and emerge as the spiritual embodiment of God's Spiritual DNA. In addition, two of my groundbreaking factors: "Bi-Mass/ Velocity Factor," that determines our fate and "Unification Proclamation: Human Consciousness" are included. The successful elimination and avoidance of tumultuous influences allows the liberated to replace chaos with an inner environment of harmony, prosperity, joy, and unprecedented freedom.

ACKNOWLEDGMENT

Without a doubt when I pay homage in this book, it must be directed to God and the place in the universe that He selected for me, providing me with my mother and father. As with everything else that I seek when unveiling the truth, I must go to the origin. My mother and father lived well beyond the average life expectancy, 88 and 91, respectively. During those years, my father instilled in me a strong sense of pragmatism. My mother empowered me with compassion.

As a result of these two forces co-existing within me, I was able to view reality with a high degree of logic and belief that, I could sense the inner feelings and needs of other human beings. It has been as a result of this influence that I feel blessed to be able to communicate my knowledge, experience, and research to humankind.

I also wish to thank all the contributing scientists, teachers, and healers who have been on the perimeter of their chosen fields. It is through their consciousness of new discovery and enlightenment that our world will remove the veil of ignorance from all of our thoughts and actions.

CHAPTER 1

THE SPIRITUAL BIG BANG: ORIGIN OF

THE UNIVERSE AND HUMAN CONSCIOUSNESS

The philosophy of one century is the common sense of the next.

- Henry Ward Beecher

The Epiphany: The Spiritual DNA Factor and Origin of Life

In order to fully understand the laws of the universe, it is necessary to be introduced to the science that governs it. For some, this is going require a revision of how you perceive the creation of the universe and its system of operation. Here, you are taken you into the mechanics of The Laws. This knowledge has existed since the beginning of creation. Just as gravity was always there, waiting for Newton to discover it, so it is with this frequency of knowledge coming through you. When you understand the structure, it gives you more data to supply the subconscious mind hologram (Explained later, and easy to understand), which will then supply you with more functional data to send to your brain for application. You will begin to think about this knowledge using your creative imagination, and then, when you discuss it with others, formulate more relevant thoughts and creativity.

In my search for answers for forty years in the field of metaphysics, it was necessary to analyze the action of Nature as we experience it in the physical world, on a moment-to-moment basis and then deduce it to the inner dimensions of existence, to the point of origin.

It is at this juncture that post modernism in the science field lacks that remaining piece of discipline, accepting the existence of an absolute, irreducible, inner, spiritual fabric of creative intelligence that is responsible for the vibratory formation of all physical world phenomena. It is for this reason that science alone will not find a viable answer to how the universe was created.

Thus far, the science hierarchy has supplied us with theories that are totally incompatible with the belief that the universe has an origin connected with a creative force. Their contradictions come as a result of their atheistic/agnostic beliefs, besides inadequate theories. It is the time in our evolution to make a changing of the guard. We need a new paradigm, a practical application that people can make sense of and find valuable in their daily lives. I believe that at some point in the twenty-first century, there will be a reunification of science and religion that has not occurred in five hundred years.

The status quo in the field of science is the Big Bang Theory, a universe created from a point called a singularity. According to theorists, a singularity is a small point within a black hole, containing a tremendous amount of matter, so much so, that light cannot escape. The claim is that the singularity became so dense with matter that it had to release the matter, so it exploded outward from the black hole, creating our universe. The problem with this theory is that it does not answer the question: From where did the matter that comprised the singularity come from? The matter that is absorbed has to come from somewhere. With the Big Bang theory, it means that there had to be a universe that existed previously in order for there to be a supply of matter for the singularity to accumulate.

A singularity is a point of infinite density and a temperature of near absolute zero. A black hole of one solar mass is thought to be at about 60 Nano kelvins. Larger mass black holes would be even colder. Hawking's theory states that the singularity exploded out of the black hole and created our universe.

Having a temperature of nearly absolute zero and then exploding to create our universe (Scientists say the temperature of our universe at the time of creation was one billion degrees kelvin.) is in direct contradiction. The laws of thermodynamics do not permit a greater density to interact into a space of lesser density. Only an entity of zero density (Light) could extend itself spatially into a void and create our universe. Also, any creation needs a source, a blueprint from which it originated. Even if a singularity had created our universe, it had to have had an origin, a creator. The Hawking's theory does not call for a creator to be involved. It was for this reason, after the announcement in 2011 by Pope Benedict XVI, that the Vatican would consider the Big Bang if God was a participant that I sent my proposal to the Vatican science department for consideration. My intention was to create a catalyst for the reunification of science and ancient scripture which has not occurred for 500 years. Since then, my source within the Vatican Science Department has informed me that ongoing discussions regarding this matter are occurring, that changes take time within the hierarchy of the Church.

Let us clear the mental fog from our minds, once and for all, that we are the only creations that possess the frequency of God. Spiritual DNA is the make-up of the physical universe. Spirit is not just reserved for we human beings on this planet; It extends Itself spatially to reside within all life. It created this life, and Its spiritual, governing body is experiencing Its infinite evolution.

And so it is with this knowledge that we can unplug ourselves from the deception that, we are the only creations in the universe that are evolving to fulfill a Divine Plan. If we were to be the main attraction in this entire creation, God would not have waited 4.5 billion years for the Earth to evolve, to create us. No, this universal evolution is a lengthy process, proceeding one moment, one action at a time. God's spiritual evolution has caused the extension of Spiritual DNA, spatially, to form this universe and then proceed on Its quest of a master plan.

I constructed the following diagram to better reveal the structure of the different levels of dimensional existence that Nature provides. At the moment of creation, the force of God spatially extended Spiritual DNA to create the fabric of space, a grid, and frequencies that carry the <u>codes</u> for the formation of all physical world forms – sub-atomic, atoms, and third dimension forms. Waves created by the <u>coded frequencies</u> vibrated over a great period of time to form particles of matter that, in turn, developed all forms in the universe.

The Wave Structure of Matter is in direct concurrence with my location of God consciousness; the wave/frequency aspect of Nature yields the mental/-consciousness; the particle/matter aspect of Nature yields the material. I applied this phenomenon to the God consciousness/human consciousness connection, taking this structure a step further and explaining the location of God consciousness existing in a spiritual, infinite, irreducible, dimension that spatially extended Spiritual DNA consciousness through frequency (Wave), which then vibrated to form density particles, creating the three dimensions of the physical universe and the origin of human conscious-ness. The frequencies carry the codes of God consciousness to human consciousness via the physical brain and the subconscious mind hologram.

Heavy density waves/particles Light density waves/particles

< God's Spiritual Dimension

Outer Physical/3rd dimension   Atomic/Sub-atomic particles

Later in this chapter, evidence, with the assistance of quantum physics, is revealed how Spiritual DNA is the essence of all physical world phenomena and how it is the origin of our consciousness. This is the Occam's razor (All hypotheses considered, the simplest one tends to be the correct one.) of explanations of how the universe was created and the origin of human consciousness.

The force of God placed all rules and regulations into life at the beginning of creation through Spiritual DNA, and in doing so, everything that began evolving from the first moment of the universe's existence has the aim or purpose to potentially extend life, infinitely. And who is to say that Spiritual

DNA has not vibrated beyond our third dimension and extended Itself with a slightly heavier density and create a fourth dimension? Perhaps, an instrument to view into outer dimensions? A macroscope would open up an entirely new dimension of science. However, when we look at the sequence of density increase as shown in the above diagram from the light density sub-atomic particles to the atomic dimension density and then to the third dimension and the forms that surround us, it would be logical to think that the fourth dimension would have to be of greater density than our third dimension, making objects of the same size as the third dimension much heavier; the particles would be compacted together to a greater degree.

God's spirit exists through all levels of time, space and causation and is the essence for all forms in the physical universe. Spiritual DNA created each dimensional level as It systematically extended Itself spatially, perpetuated by Its own vibratory frequency activity, just as our own DNA is the essence for every aspect of our human makeup. As we look at the illustration of the external and internal structure of the universe, it is clear that the outer physical world (3rd Dimension) as we know it can be perceived with our five senses. However, there exists levels of creation that cannot be perceived by our five senses; these are the sub-atomic and atomic structures.

We know that in order for all forms we see around us to exist, the atomic elements of electrons, protons and molecules must be present. When the universe was created, Spiritual DNA formed the fabric of space and a grid to carry the flow of coded, spiritual frequency into all creation. Eventually, vibrating waves appeared and manifested particles of matter. As stated above, each dimension level (Sub-atomic, atomic, and 3rd dimension) was created, and each level vibrates with a different degree of density. The density intensifies as the vibration travels from the inner dimension to each successive dimension. Our 3rd dimension vibrates with the heaviest density.

We can take the explanation of our inward journey one step farther. To insert an analogy, suppose we want to construct a building. We would need some type of drawing or blueprint to give us an idea of how the building would look and serve as a foundation for the physical structure. So it is with the external structure of the universe. We need to have an underlying foundation for all that we are experiencing in our everyday lives. This is the spiritual existence of God that Jesus Christ taught two thousand years ago and which was known by the ancients of the East six thousand years ago, Buddha in 6th Century B.C. and Muhammad, 6th Century A.D.

This may all seem to be difficult to comprehend at first; however, with some logical thought, the abstract can become very concrete. You might say the elements of creation are "hiding in plain sight." From some religious observations, the placement of God within everything may be held as blasphemous.

Let us look a bit closer at the idea that God is in everything. True, the force of God does occupy God's own infinite, irreducible, internal, spiritual dimension; however, God extended Spiritual DNA, spatially, to create our universe - God's off-spring. This is the "Grand Paradox," the one, infinite consciousness also existing as the many. In order for us to be created in the image (Spirit) of God, we certainly have to be connected through Spiritual DNA frequency, not separated. Therefore, God must dwell within us as spirit, at the inner origin our physical body.

As with all God's creations, God's spirit must be a vibrating frequency that connects within us and allows our physical body to function in the 3rd dimension. As a result of this interconnectedness, God establishes an omnipresence that exists throughout all creation. Omnipresence within all physical world structure carries with it all the laws of God. The laws of Nature were set into place and activated at the time of creation. That is the reason why our actions will automatically receive certain retribution. No action goes unnoticed. God's law is on automatic pilot, vibrating within everything. This is the basis for the Law of Karma.

Everything in the physical universe is an extension outward of God's Spiritual DNA. God's spirit manifests outward (By means of my discovery of Vector Causation) through time, space and causation, outward into the sub-atomic and atomic realms, outward through the invisible cellular realm and into the realm of our five senses, the realm of physical objects that surround us.

With this knowledge in place, it is clear that Spiritual DNA originated all life with Its one vibratory action. To understand the works of God, the works of Nature, this is our true quest as human beings. When we discover our true potential, then will each of us be able to serve God and feel our inter-connectedness with Spiritual DNA. You must go within yourself and hear the silence of God's voice. It is only then, that you can hear your calling from God, and then express your soul to the world. Let us bury the archaic brain conditionings of the world that discriminate and advocate inequality because of race, color, and creed. All humanity vibrates with the same Spiritual DNA frequency. It is only our perception of reality that creates static in the receiving station of our brain.

The Illusion is Within

When you experience a shift in this perception, you begin to realize that the only misery that exists is your belief system about life. Becoming aware of this, you can begin to challenge the beliefs that for so long held your mind in bondage and prevented it from actualizing the truth that, your potential resides within.

You will begin to enjoy life more from a spiritual consciousness, rather than the mental and emotional levels that provide you with a constant supply of pleasure, followed by misery, followed by pleasure, followed by misery, etc., etc., etc. This is our worldly experience in a nutshell; we perceive life with our five senses and interpret this life based on environmental stimulus. We can never be truly happy when we perceive reality from this endless flow of change created by Nature's flux. This is the grand paradox; we believe God to be an immutable, almighty, spiritual consciousness protruding through time, space, and causation but creating a world that is in direct contrast to that which is divine. It is our conditioned brain that perceives creation to be an illusion.

With this concept in mind, it is easy to realize that creation has God deep within it. From this we can now begin to apply a new philosophy to our daily lives. No longer must we treat only that which we like with love and respect but all creations, both animate and inanimate. This is the philosophy that all the great prophets taught, the vibration of God's love that manifests upward through all the dimensions of existence. This love vibration is the glue that holds the universe together. When you realize this great truth, you can then begin to express gratitude to God for every encounter that you experience, knowing that it is God's force that will manifest into physical form all that you need. The two, great, universal laws that provide this phenomenon are: the Law of Association and the Law of Repetition. Later, I will explain the science behind these universal truths when I discuss the subconscious mind hologram.

This will create a new sense of faith in your life by lifting the negative, emotional burden from the way in which you react to life's situations and circumstances on a daily basis. This is a life changer! When you experience this dramatic shift, you will never desire to return to your former path of behavior. I think everyone has seen the movie, "Pay It Forward." Tomorrow, if everyone in the world had a change of consciousness that, "Today, I am going to give to everyone I encounter more than they give me," the world would never again want for anything. Can you imagine arguing with another person over who is going to give more to the other, rather than receive? This is love manifested to its fullest potential. This is realization that God exists in all things.

To better explain the workings of the underpinnings of universal structure, the following reveals evidence and will help you visualize how we receive the origin and nature of our consciousness.

UNIFICATION PROCLAMATION: HUMAN CONSCIOUSNESS

By

Dr. Damon Sprock

The purpose of this writing is intended to reveal the interconnectedness between all the components of existing knowledge regarding human consciousness, a system containing an origin, function, and insertion into all universal activity. To achieve this challenge, it is necessary to begin at the source.

In my search for answers, it was necessary to analyze the action of human consciousness as we experience it in the physical world on a moment-to-moment basis and then deduce it to the inner realms of existence, to a point of origin. It is at this juncture that, post modernism, in the science field, lacks that remaining piece of discipline, accepting the existence of an absolute, inner fabric of creative intelligence that is responsible for the vibratory formation of all physical world phenomena.

Spirit is the origin, as metaphysics teaches. It is the cause of all reality and thus the solution to how all creation is interconnected. If science can accept that the universe was created from a Big Bang, with a physical universe instantly coming into existence from either nothing or a singularity within a black hole, then is it not conceivable for science to shift and evolve to a potential consciousness, the belief that a creative intelligence existing within all physical world levels could be the point of origin? Could it not be conceivable that the premise that, there must be something beyond <u>that</u> fabric of consciousness in order for <u>It</u> to exist be invalid? (Irreducible Factor) Would it not be potentially logical for there to be a dimension consisting of an infinite, spiritual, fabric of internal consciousness that is irreducible and that is responsible for reality?

The Pure Light of spiritual consciousness contained the potential to extend Itself outward (Vector Causation), creating the fabric of space and as Einstein proposed, extend spatially. Spatially extended, vibratory, Spiritual DNA energy gave rise to all physical forms that we experience with our five senses.

In essence, what we can perceive as reality is a 3D, external, infinite, physical universe consisting of matter and supported and balanced by an internal, infinite, irreducible, dimension of spiritual consciousness. This can be visually observed as a balancing see-saw with a fulcrum at its center. In my scenario of universal structure, the internal and external worlds are balanced and bridged at a fulcrum point I refer to as "Spiritual-Matter Transference."

3ʳᵈ Dimension   Atomic   Sub-Atomic

.(.(.(.(.(.(.(.(.(.(<<<<<<<Infinite   Spiritual   Consciousness

_____

"As inward, so outward"

Vector Causation

^

Fulcrum of the "Spiritual-Matter Transference Point"

It is at this juncture that creative Spiritual Consciousness is transformed into matter, the physical world. This is the very essence of Creative Consciousness in the form of Spiritual DNA. Spiritual DNA spatially vibrates outward (My discovery of Vector Causation, "As above, so below," is revised to, "As inward, so outward.") and becomes the essence of the physical world. It contains all the laws and regulations that govern every action of life, therefore, giving It omnipresence, an authority over all things.

Dark Matter

One of the problems facing scientists is the inability to determine what the

"Dark Matter" (Absence of light) between physical world forms should be termed. There is no mystery if one understands the action of "Spiritual-Matter Transference."

The "Dark Matter" is also Spiritual DNA. It is the frequency that vibrates throughout all three dimensions (Sub-atomic, atomic, 3rd dimension of space). It is the omnipresent, underlying, base level of Spirit that allows space to exist and thus spatially extended matter and thus motion and time.

Spirit is the origin, as metaphysics teaches. It is the cause of all reality and thus the solution to how all creation is interconnected. This is the basic principle of all origin and insertion principles, origin being Spirit and insertion being continuous space with spherical waves and particles continuously vibrating from the <u>coded</u> frequencies the substance of life throughout every form in the universe.

In order for Spiritual DNA action to occur, it is essential for Its vibratory flow to establish a pathway for Itself from Its origin to Its insertion. As we scientifically observed, matter consists of tiny particles of sub-atomic life. The scientific findings of physicist Geoff Haselhurst in his extensive expansion of the "Wave Structure of Matter Theory," as proposed by Erwin Schrödinger and Louis DeBroglie (1927), describes matter, "As being spherical waves in continuous space, reality described as resulting from One thing, space."[2]. This is in direct acceptance with my Unified Theory. In addition, I extended the "Wave Structure of Matter Theory" to include its relevance into the realm of human consciousness, where it vibrates all preexisting potential into our subconscious mind hologram.

First of all, the distance between God's infinite dimension and the three dimensions of the physical universe is minuscule in nature. We can see the inner atomic and sub-atomic dimensions with a microscope. God's spiritual realm is inward from those two dimensions, so close. The misconception occurs when archaic beliefs of God existing outward in the heavens are mainstream acceptance without science entering the equation. When we observe the entire universe outward from our view here on Earth, we can say that God exists out there; however, you must realize that the whole of the universe, no matter what part of the universe you go to, will have just within it the three physical world dimensions at that location in space (Sub-atomic, atomic, 3rd dimension) - God's spiritual dimension.

This is a term called non-centrality; there is no central point in an infinite universe; any point could be considered the center. The paradox that exists here is that, although there is non-centrality with regard to no place in the physical universe being the center and God's infinite, spiritual dimension existing within it, centrality of God's infinite, spiritual dimension is definite in that It is centrally located at the spiritual dimension within all physical world phenomenon, no matter what part of the universe you may go. This also explains that wormholes leading to other parts of the universe do not and cannot exist. Anything of density cannot pass through a section of the universe and come out at another part. It is not possible to enter God's infinite, irreducible, zero density, spiritual dimension of pure light and then exit at another point in the universe. The journey dead ends at God's spiritual dimension.

The following diagram illustrates how this phenomenon functions:

Frequencies Permeating the Physical Universe

4th Dimension |||||||||||||HELL||||||||||||||||||||||||||||||||||||||||||||||||||||||||||||||||||||||||||||||||||||

|||||||||||||||||||||||||||||||||||||||||||||||||||||||||||||||||||||||||||||||||||||||||||||||||||||||||||||||||||||||||||

3rd Dimension ||||||||||||||||||||||||||||||||||||||||||||||||||||||||||||||||||||||||||||||||||||||||||||||||||||||||||||||||||||||||

|||||||||||||||||||||||||||||||||||||||||||||||||||||||||||||||||||||||||||||||||||||||||||||||||||||||||||||||||||||||||||

Atom Dimension ||||||||||||||||||||||||| Frequencies carry <u>codes</u> |||||||||||||||||||||||||||||||||||||||||
for all physical world forms.

Sub-atomic Dimension ||||||||||||||||||||||||||||||||||||||||||||||||||||||||||||||||||||||||||||||||||||||||||||||||||

|||||||||||||||||||||||||||||||||||||||||||||||||||||||||||||||||||||||||||||||||||||||||||||||||||||||||||||||||||||||||||

_____Fabric _____ ^	_____ Baseline fabric of the Universe	_____ ^ _____Fabric_____
Human spirit vibrates at this transference point.	"Spiritual-Matter Transference" "As inward, so outward" God's Spiritual Dimension	Pure Light becomes matter, as frequencies vibrate through physical dimensions.

The fabric of the universe begins at the baseline of Spiritual-Matter Transference.

This is where you must really focus your attention on dimension. When I speak of God's infinite, irreducible, spiritual dimension, you have to imagine this dimension existing infinitely in all directions and then spatially extending Itself infinitely in all directions (My discovery of Vector Causation), creating our three physical universe dimensions. For as infinite as God's spiritual dimension is, that is how infinite the outer, three dimensions have to be in order to receive support. Remember, God's dimension is situated just within the physical world's three dimensions. No matter where you travel in space, God's spiritual dimension is vibrating with frequency at the baseline of that part of space. This is how God's Law of Omnipresence functions.

To further understand this phenomenon of universal dimension, an analogy is needed that is familiar to us all. Everyone has used a flashlight. Suppose we have a flashlight with three settings. The battery and the filament would be the origin, the source of the light beam. However, let's not limit the flashlight's source as just a single, small point. Let's visualize the source of the flashlight's power as stretching in all directions (Vector Causation) into infinity, occupying its own, infinite dimension of origin.

With this in mind, it is now possible to envision how all dimensions created from the source will possess their own infinite dimension. When the flashlight is turned to the first position, the frequency of the light extends outward in all directions (Vector Causation), to a certain point, a short distance, because the <u>density</u> of the frequency is lightest closest to its source. This is analogous to the sub-atomic particles having the lightest density and being closest to God's source. When the second position is selected, the frequency of the light is extended farther, because the light <u>density</u> is heavier – analogous to the atomic 2nd dimension. And when the third selection is chosen, the light beam reaches a greater distance – analogous to the third dimension of the universe. Each position of the flashlight extends the light to a certain distance; each has its own dimension of <u>light density</u>, yet all three dimensions of light have the same <u>origin</u> of frequency – the flashlight. Without the origin, the outer three dimensions do not exist.

Now, what happens if you turn the switch to the "off" position? The light returns to its source. If the force of God decided to "flip the switch" and discontinue creation, the underlying, quantum frequencies that support the three physical world dimensions would collapse in an instant and return to God's source, a reverse Big Bang.

This energy grid changes its frequency over the millenniums, creating a change in the manner in which we perceive reality. This is the manner in which we evolve, consciously, to expand our knowledge as a species. Our evolving is not a result of chaotic randomness. No, there is a design to all of this. We, as a collective force of moving mass, create our destiny. We have the power to move this huge mass of humanity in a direction that will raise the level of frequency we encounter on a moment-to-moment basis and thus produce levels of consciousness conducive to better decision-making, raising the percentage of positive results occurring in our lives.

Frequency in our energy grid can be likened to the frequency we encounter on our radios. When the tuner is brought closer to the center of the power source, the clarity of the music is brought to its fullest content. The energy grid is constantly fine tuning our consciousness as humans, eventually bringing the clarity we need to end mental and emotional misery as we know it. The following work is entitled, "Bi-Mass/Velocity Factor," which explains how the collective forces of our continuous momentum as humans have a direct affect upon our destiny.

"Bi-Mass/Velocity Factor"

Dr. Damon Sprock

Random: Webster's Dictionary assigns the definition to random as: without aim or purpose. We have people in the twenty-first century who still believe that the whole of life is based upon random events. There is no room for this philosophy in our world. Stop using "random" when describing occurring events. Random is an excuse for not taking responsibility for one's actions. Everybody is right; nobody is wrong. I first recognized this fallacy with the introduction of "No fault" car insurance: accidents are a result of random events. No, accidents are the result of poor, human decision-making: text messaging while driving, speeding, under the influence, etc.

Everything in the physical world has a nature to it, animate and inanimate alike. Every particle of existence has its own, unique way of performance. All particles of matter, large and small, have their own role to play in the "wheel of life." These duties are performed in accordance with the natural laws that were established at the time of their creation. All things must adhere to the rules of their individual makeup, or perish. Every part of the "wheel of life" is necessary for every other part to exist. To believe that all events are attributed to randomness is absurd. The whole point of this universe is life. In order for life to continue, orderliness must exist in all things.

In randomness, chaos would be "key," and even chaos would have to have its own nature. It would have an order to follow. It would appear that chaos had no measured form or purpose; however, it would have a purpose, to be a form with its own nature - disruption. The universe itself is thought by scientists to have been formless matter at the beginning of creation. The simple fact that, we are where we are at this very moment is proof in itself that all events that occurred previously were necessary and with purpose.

At this point, it is necessary to establish concrete validity for my factor. I shall achieve this by utilizing basic laws of physics. Substituting people in place of inanimate objects, two people traveling in a direction toward one another, each at his/her own speed, will eventually collide (Human interactions in this case) due to each other's personal, inner desires (Thoughts, emotions, desires). Paths cross, a scenario occurs, and the two people swap momentum. The masses (The bodies of the two people) remain the same; however, the velocity has changed (The bodies have stopped.). In the equation, $P=MV$, P is the momentum of the two people and is determined by the mass (M) of the people times their velocity (V).

Momentum plays a dual role in all space-time activity. Momentum is continuous in all events, first playing the role of cause as it moves in any direction and creates the moment, which then becomes the past with each passing moment and then unfolding into the effect, creating the future. However, momentum never really <u>acts</u> in the past or in the future; its movement only experiences the present. The past and future only exist as a result of momentum continuously moving in the present from one point to another. The stick people will function as our roles in the action of momentum.

Before A and B collide, each was involved in a previous encounter and then eventually became involved in a future encounter with C and D, respectively. The previous engagements, as are the present and the future, were made possible by each person's desires under the influence of their thoughts and emotions occurring in a moment-to-moment continuum.

We decide (Free will), and we are responsible for what our next momentum is going to be. There are no accidents, only decision-making. When two forces collide and interact, the person with the lesser engagement proficiency (Weaker of the two) will be directed elsewhere while at the same time creating a new momentum, P=MV, toward another person who is establishing the same following his or her previous encounter. The phenomenon of synchronicity can be explained utilizing this factor, similar events occurring with no related causal connection can be attributed to the "Equation of Human Events." The following illustrates the continuous dual role of momentum (E), E representing P as energy whereas:

$$E = \rightarrow \overline{MV}^{2} \rightarrow \text{The Bi-Mass/Velocity Factor Equation}$$

A+C
Forces Collide, Swap Momentum

Vector C

O

(MxV)
C

A+B
Forces Collide, Swap Momentum

Vector A   **Vector B**

O   O

(MxV)   **(MxV)**
A   **B**

B+D
Forces Collide,
Swap Momentum

Vector D

O

(MxV)
D

During football games, the announcer will mention how the momentum has changed and is now favoring one of the teams. When one takes into account the actions of twenty-two players on the field and the possible number of interactions that can result due to the moment-to-moment portrayals of each player's belief system as a result of brain conditioning, it is now clear to understand how just one misdirected action by a player can change the course of destiny. Any number of X factors can manifest at any moment.

This continuous momentum of human objects being propelled toward one another to engage interactively and then projected toward other human beings who are experiencing the same action creates what I term the, "Equation of Human Events," which continues in an infinite manner and moved through space-time by the chosen thoughts of each participating individual. These choices are made as a result of previous choices based on reason, intellect, and emotion directed toward some aim or purpose. This is the key point for the proof that randomness in events is not a factor; meaningful and willing decisions are made based on the belief system and brain conditioning of human beings, which are continuously unfolding due to what has occurred previously.

With this material in mind, it is now possible for human beings to comprehend the reality that every interaction with another human being has a meaningful purpose, and that their existence and momentum is necessary in order for all other action to unfold. This single factor established into the psyche of the human element casts a new light and a shift in human consciousness, the light being our acceptance of all that happens to us on a moment-to-moment basis, whether that incident concurs with our desirous plans, or not.

The "Bi-Mass/Velocity Factor" explains everyday phenomena that occur in every human being's life. Scientifically understanding why things happen in our daily lives rids our minds of many negative thoughts and emotions. Occurrences that once caused us to become angry, blameful, resentful, jealous, etc. can now be viewed with clear perspective. I assure you, that when you begin to accept all that happens to you, your luck will change by Divine Decree.

Knowledge of the intricacies of the "Bi-Mass/Velocity Factor" allows you to achieve a shift in consciousness that creates clear insights into the phenomena of our universe. Just as a shift in the value of a currency can accumulate vast rewards for investors, the same holds true when mass acceptance in society occurs regarding a new paradigm, sending human actions in directions that accumulate rich rewards in the arena of human relations.

Time

As we now observe how momentum plays a dual role in the action of all events, it is at this point that we must realize that time is an illusion. Time was established by humans to measure the action of momentum from one point to another. As Geoff Haselhurst describes, "Time is caused by wave motion."(As a spherical wave motion of space which causes matter's activity and the phenomena of time.) 3. If we lived in another part of the universe where it was either total darkness or total light, or, we had no sun to travel around from which we could measure the length of a year, month, week, day, hour, minute, and second, how would we measure anything? There is only the existence of the present with the action of momentum occurring. The "Bi-Mass/Velocity Factor" proves this. There is only momentum moving continuously in the present. This can be illustrated with a ball rolling on the ground from point A to point B.

A   O>>>>>>   B

As the ball's momentum propels forward, its reality is always in a state of experiencing the present; it is never experiencing the past every moment, nor is it experiencing the future every moment. Its momentum keeps it in constant present. No momentum travels from its past to its future without experiencing present.

An example of proof that the present is all that exists is when you are driving your car. If you are driving on a road and a distant view can be observed, you may say that the view is in your future as you move toward it; however, to a person who is already situated at that distant location, it is that person's present. The same concept concurs with the distance you have already traveled. It may appear as your past; however, it is the present for someone behind you. Therefore, everything is a present situation; we only appear to change that fact when we propel our momentum in any direction. The physical universe is an infinite construct, an extension of inner, spiritual consciousness containing a set of axioms that apply to all working systems.

Holographic Psychology

The next phase of interconnectedness among all systems that establish human consciousness is the essential work of my late colleague, Dr. James Pottenger, renowned researcher in Science of Mind and protégé of the late Dr. Ernest Holmes. Dr. Pottenger was mentioned previously in this writing.

Dr. Pottenger's system relates to paradigm shifts within human consciousness, bringing into existence new levels of understanding and altering an individual's view of reality. The newly established paradigm creates a shift and exposes three different levels of comprehension in which humans express their view of reality. Holographic Psychology presents a new paradigm and explains the need for shifting from external, objective, world stimulus to internal, subjective, spiritual acceptance. It is where the individual discovers the source of all existence as a collective base that is innate in Nature.

Holographic Psychology explains three levels in which humans perceive reality. First level involves action in our perception using emotional content. This is a result of our five senses interpreting observations. Many people react to everyday situations only with emotions. This can be harmful. Not having a self-awareness of the consequences of emotional reaction can result in experiencing much misery.

The second stage of development used to interpret the environment is self-awareness. When an individual is able to remain calm while experiencing daily activities, he or she can choose to replace it with a more acceptable conditioning. It is when humans become more aware of whom they are and responds to the environmental stimulus with more understanding of the nature of things and why they happen is when a shift in perception occurs and allows one to act out his or her decisions with a higher degree of consciousness.

"Learning is realized as inner acceptances, rather than a result of external causes. The third and most profound level reveals the true essence of who we are and our relation and interconnectedness to all things and the spirit of creation regarding an all Universal Mind. Third reality functions as one's source, the question of epistemology or philosophy of the nature of knowledge awakens to a total change from our early realities. We discover that, how we know and what we know is an intuitive awakening.

This paradigm shift in understanding entails a preexisting potential that is omnipotent (All power), omniscient (All intelligence), and omnipresent (Everywhere present). It is one of the most important discoveries in the late twentieth century, revealing that the self-image is part of a process of human perception that evolves through three different realities. It is a realization that the senses do not give us meaning and feeling regarding our so-called external world. What we "see" (Sense) externally from us is not the cause of our frustration but the meaning we ascribe to that sensed experience."4.

Once again, throughout the entire gamut of the physical universe extends an underlying fabric of spiritual consciousness, Spiritual DNA. This consciousness is the source, the origin of all our knowledge, including all potential knowledge yet to be revealed through third reality observation. So it is that, a spiritual fabric underlies all physical forms; it is therefore the force that underlies our physical brain and all that the brain controls. The pathway of each section of the brain is directly interconnected with the system of Spiritual DNA consciousness. With this understanding intact, it is now visible to acknowledge that the activity of the brain has an underlying force from which a vast potential, a reservoir of knowledge resides.

Subconscious Mind Hologram

The subconscious mind hologram is our connecting link in the Universal Mind. All minds are linked to spiritual consciousness that vibrates all existing knowledge available through spherical, frequency waves in continuous space.

How does one tap this potential knowledge? Subconscious mind hologram power is also a system. It is a system that allows us as humans to interact with the unlimited potential that exists within the dimension of spiritual consciousness. The subconscious mind hologram is that aspect of our consciousness that allows us to access the answers to the external, physical world. This occurs as a result of the spiritual fabric creating the subconscious mind hologram link through its interconnectedness with the physical universe.

The procedure of connecting with the spiritual fabric of knowledge is quite simple; however, it is important to understand certain facts concerning the subconscious mind hologram. First of all, it must be guided. Herein lays the paradox; the system containing all universal knowledge must be assisted by the finite mind of a human. So it is; however, we must follow universal rule.

The subconscious mind hologram must be instructed, just as you would search for information on the internet, using words. The only difference is that we need to accompany the words with emotion and detailed, picture images of the solutions we want to achieve. When the eyes are closed and centered to the inside of the forehead, the frontal lobe, the visual imagery that you create is sent to the brain for review. The imagery is analyzed and then sent to the great vault of potentiality, the subconscious mind hologram, where it collects the necessary data to resolve the picture you have just sent it. The data is then relayed back to the conscious level of brain activity. During daily activities, the answers to the visualization flow into the decision-making process of your brain as momentous action. The swiftness of arrival of the answers is subjective and is in direct proportion to your receptivity of subconscious mind hologram principles, the result of your belief system due to brain conditioning. The action of computers follows the nature of our brain and subconscious mind hologram. The brain is the hardware that supplies the subconscious software with search data. The subconscious software's function is storage and translation.

This takes us to the final destination, the actual location of human consciousness. The subconscious mind hologram is the hub of our consciousness; however, there are sub-divisions that need mentioning. Every part of our body contains living elements of our physical being. Every cell is a living entity and has, as a part of its makeup, a memory of its actions. The central nervous system with its vast network of pathways, featuring data transferred via motor neurons, contains a motherboard of intricate connections from the internal grid of our being. Information is continuously vibrating from all points of our physical being through our cellular structure that comprises all living tissue. The total access of all memory throughout our system constitutes what is known as consciousness. This interconnectedness among all memory-capable systems and the brain creates consciousness during wake states, hence the conscious state of mind.

Every part, every minute portion of our body consists of cellular structure, billions of tiny cells that comprise tissue, organs, vascular system, and nerves. In total, the simultaneous vibration of this entire network, every moment we are alive, constitutes the total gamut of information that our mind contains in its observation of reality. As the vibrating frequency of our cellular structure spatially extends to form our dense body, with the internal, spiritual realm functioning as its source, so it is the invisible fabric of spirit extends spatially through every level of our body and all other forms in the physical world.

An additional spatial extension that our consciousness vibrates to within our being are the seven centers of consciousness, page 26. This system has been acknowledged for thousands of years; the ancient monks of the East devoted their entire existence to observing Nature and how we co-exist with its force.

As the following illustration indicates, five of the centers are located along the spinal plane and two at the cranial level. Each center has its own function, and each contains an inner mental body, an intermediate energy body, and an outer physical body.

The first three centers constitute the lower mind, and they are: the base of the spine, the genitals, and the solar plexus. Through the base of the spine are activated primitive instincts, self-preservation being the basic ingredient. The genital area expresses the human sex drive. The solar plexus initiates digestion and assimilation of food into the bloodstream. It also activates energy to perform work. The heart center begins the transition from the lower mind to upper levels of consciousness. It is the fulcrum, the balancing center.

The last, three centers are considered the upper mind, and they are: the throat, the third eye (Frontal lobe), and the crown. The throat is recognized as the power of the spoken word through which we communicate. The third eye is responsible for the visual imagery that is transmitted to the brain and then to the subconscious mind hologram. The crown center is the highest level and represents higher consciousness. The psyche consists of two levels of self- "False Ego" and "True Self." "False Ego" may be eliminated and "True Self" discovered at this level. "False Ego" can be eliminated by visualizing unwanted behaviors exiting through the top of the head and then evaporating.

Every aspect of our consciousness can be entered and explored through this system of centers. The potential lies within each center and contains intercon-nectedness to all aspects of life, and if pursued in the proper manner, it will enhance the quality of life of all that it is connected to.

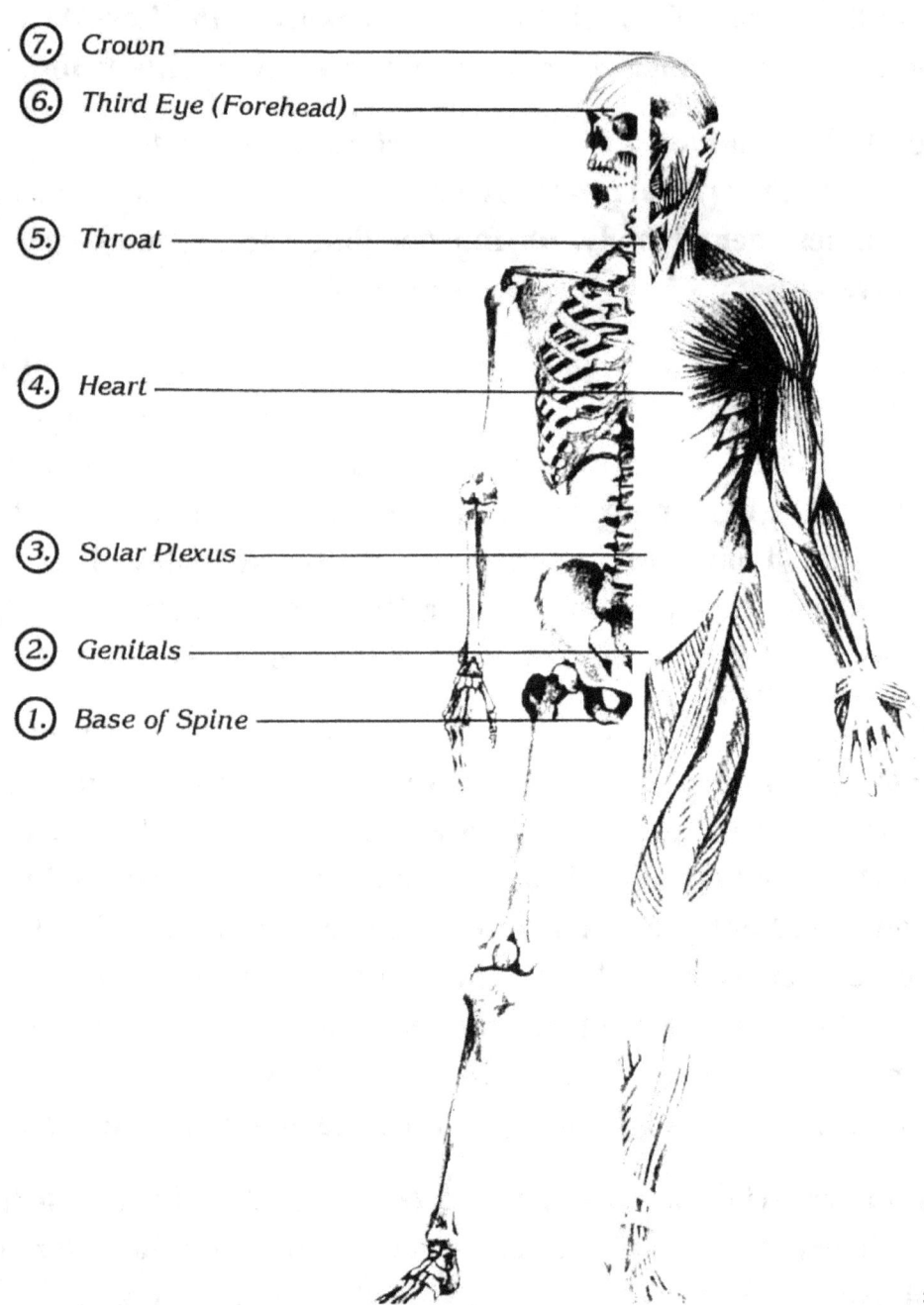

⑦ Crown

⑥ Third Eye (Forehead)

⑤ Throat

④ Heart

③ Solar Plexus

② Genitals

① Base of Spine

5.

7 Sub-Centers of Human Consciousness

The Quantum Connection to God Consciousness

We have now reached the final phase of our search for the missing link. What is consciousness, and where is its origin? As you have previously discovered, the universe extended outward, spatially (Vector Causation), from God's internal, infinite, irreducible, spiritual dimension. The fabric of space was created along with a universal grid that carries the <u>coded</u> frequencies for all physical world forms. The frequencies hold all pre-existing knowledge that emanates from God's spiritual dimension. When we educate (The word educate originates from the Greek word *edu* that means, to flow outward from within.) ourselves, we are bringing forth pre-existing knowledge from the universal frequencies.

This is a unique process that the ingenious consciousness of God has provided. We have within our conscious mind a quantum, subconscious mind hologram storehouse of knowledge. This vault is extremely vast and contains all the data available in the physical universe as a result of its connection to the universal grid of frequencies of God consciousness that permeates our entire existence. To gain control of the subconscious mind hologram memory bank is to enhance your mental capacity, immensely.

Again, the subconscious mind hologram can be likened to a computer. The brain is the hardware and the subconscious mind hologram is the software, the brain's underlying, mental plane of memory. Also, we possess an added element that computers do not, emotion. So here you are, armed with an enormous amount of data and the mental/emotional energy to direct that data in a manner that can evolve you to levels of consciousness never before experienced but no road map to guide you.

Do you realize that your subconscious mind hologram is seven times more powerful than your conscious mind? You have to ask yourself, "Why am I only using 10% of my brain capacity?" Well, the answer is that you have not been introduced to a model that allows you to tap the vast storehouse of knowledge that the subconscious mind hologram contains. The conscious mind and subconscious mind hologram work together in perfect harmony.

Years ago I coined a term, "Pendulum Effect," regarding the action that takes place between these two forces. This phenomenon works as follows. The conscious mind is the rational portion of the mind, incorporating reasoning and intellect and functioning with the aid of the five senses. For instance, if we see a fire burning, we know through our conscious mind not to touch the flame. However, it is the subconscious mind hologram that stored the information for us to call upon. The subconscious mind hologram not only stores every event in our lives, it translates data needed to solve problems and aids us in achieving our goals. To demonstrate this, I will give you an example.

Suppose that you want to learn how to play tennis. One of the most important phases of tennis is the serve. Of course, you would want to obtain the services of a qualified instructor. The instructor would take you through the entire process of serving, step-by-step, until the motion of serving was learned.

Consciously, you would have to learn: how to hold the racquet, how to stand at the serving line, how to toss the ball above you, when to swing the racquet and at what angle, so that both ball and racquet meet at the right point, and then, how to follow through. Depending upon how skilled you are, you must now remember each of these steps as you begin your practice sessions. Then, after numerous repetitions, you begin to discover that there is no longer a need for you to consciously think of each step separately. What has occurred is the subconscious mind hologram has received many images of how to serve a tennis ball and has given instruction to all of your involved body parts the necessary information of how to perform a proper tennis serve.

The subconscious mind hologram is also referred to as the automatic mind. It, too, is responsible for all the involuntary body functions that we do not have to give conscious thought to such as: the heart, lungs, blood-flow, stomach, etc. This is its holographic nature - storage and translation.

There are two, dynamic laws which govern the function of the subconscious mind hologram, and they are: the Law of Association and the Law of Repetition. The Law of Association states that, the subconscious mind hologram will absorb into itself, through the assistance of the five senses and the brain, all influences from the physical world, and it will be most influenced by detailed, picture images and emotion; it is the seat of emotions. Picture images allow the subconscious mind hologram to search its memory banks more efficiently and relay the proper data outward to the brain for application. A picture tells a thousand words.

The Law of Repetition states that, placing detailed, picture images repeatedly before the subconscious mind hologram to absorb will imprint those images into the memory banks as the proper data for that specific solution.

Placing yourself in a relaxed state, both physically and mentally, is most conducive to influencing the subconscious mind hologram. This state of relaxation allows the mind to drift into alpha frequency, the level of brain rhythm most vulnerable to suggestive pictures. Since the subconscious mind hologram has no reasoning powers, its process must be initiated by outside sources, you or an assistant. The subconscious mind hologram will accept the most dominant of influences (Repetitions) sent to it. It will provide negative answers as well as positive ones, so, be mindful of the thoughts and images that you provide it.

When you close your eyes, look upward to the area between your eyebrows. Use the inside of your forehead as a picture screen. This is called the "Third

Eye" that Jesus Christ mentions in the Bible, "The light of the body is the eye; if therefore thine eye be made single, thine whole body shall be full of light."6. (Mathew 6:22). Place all of your picture images on this mental screen. The subconscious mind hologram will immediately be impressed by these images and begin a detailed search for the information you have requested, in the same manner your computer does when you are seeking subject matter on the internet. The most dominant images in repetitious form will be impressed permanently into the subconscious mind hologram memory and acknowledged as truth. The "Pendulum Effect" is created by the conscious mind sending images to the subconscious mind hologram for resolve. The subconscious mind hologram translates and then relays its answers to the conscious mind. You are impressed with the results and respond with your conscious mind by sending more images to be resolved. And so, the pendulum swings.

The two periods of the day when the subconscious mind hologram is most vulnerable are: upon waking in the morning and when retiring at night. At these two times, your mind enters into the subconscious mind hologram state by passing from sleep state to awake state and from awake state to sleep state, respectively. These are excellent opportunities to experience alpha brain rhythms and create pictures expressed with emotion (The subconscious mind hologram is the seat of emotions.) for the subconscious mind hologram to program into its data banks. Archaic, unwanted, behavior patterns can be replaced with new, workable solutions. What world do you want to live in?

The following chart depicts the different levels of vibration or frequencies that you may train your brain to experience. Each frequency allows you to achieve the results listed in the right column. Our thanks to Mr. Philip James Eastwood for his contribution of this collector chart depicting the four levels of frequency that are obtainable when relaxation/visualization techniques are diligently performed.

As you can see, when one becomes proficient at lowering the cycles of brain rhythms, it is possible to enter realms of the mind conducive to health and well-being. Each person will succeed at his and her own pace. I strongly suggest that you seek a trained instructor for guidance. Just as you would train your physical body with a qualified person, so should you do the same for your mind.

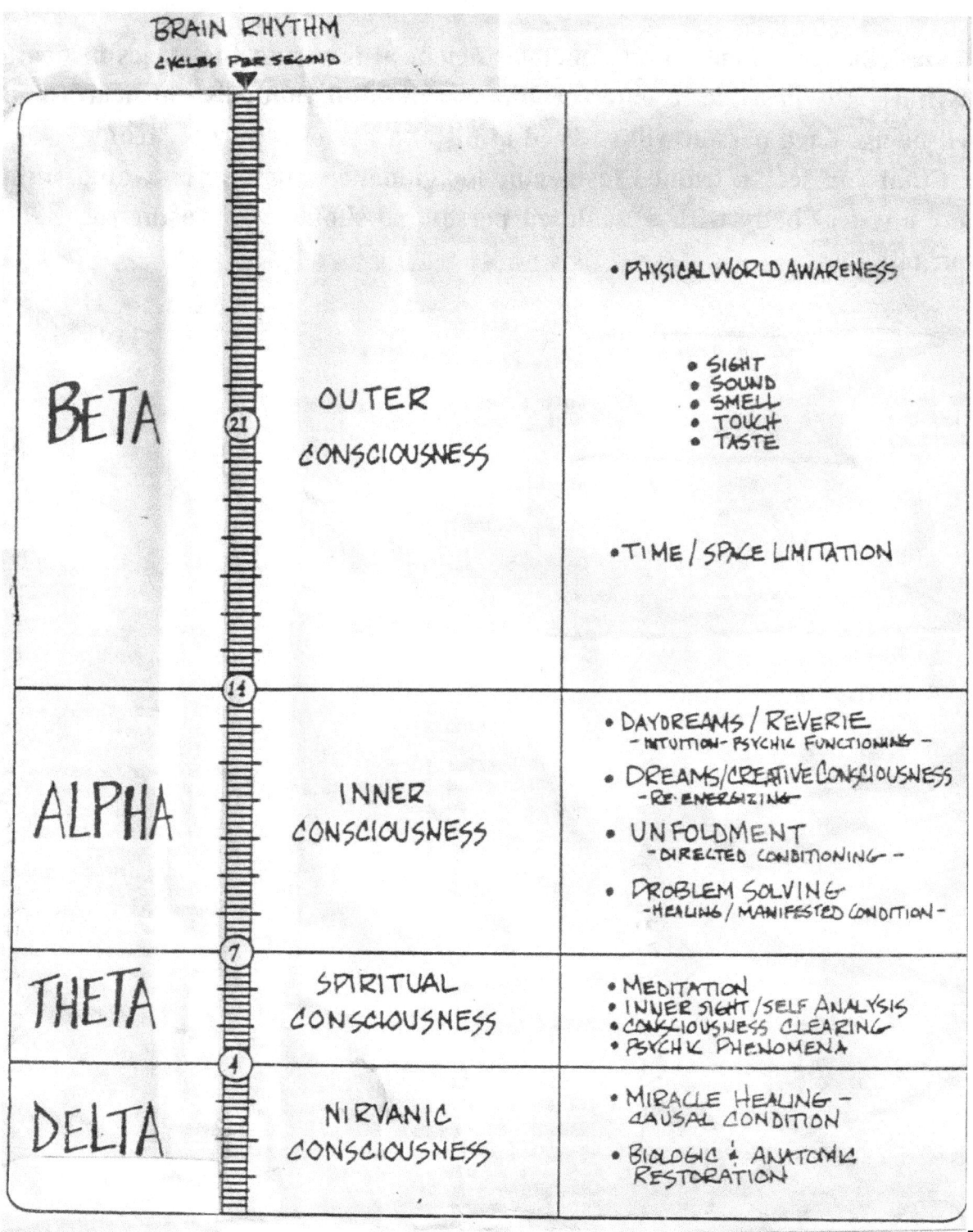

BRAIN RHYTHM
CYCLES PER SECOND

BETA — OUTER CONSCIOUSNESS
- PHYSICAL WORLD AWARENESS
  - SIGHT
  - SOUND
  - SMELL
  - TOUCH
  - TASTE
- TIME / SPACE LIMITATION

21
14

ALPHA — INNER CONSCIOUSNESS
- DAYDREAMS / REVERIE
  - INTUITION- PSYCHIC FUNCTIONING -
- DREAMS / CREATIVE CONSCIOUSNESS
  - RE-ENERGIZING -
- UNFOLDMENT
  - DIRECTED CONDITIONING -
- PROBLEM SOLVING
  - HEALING / MANIFESTED CONDITION -

7

THETA — SPIRITUAL CONSCIOUSNESS
- MEDITATION
- INNER SIGHT / SELF ANALYSIS
- CONSCIOUSNESS CLEARING
- PSYCHIC PHENOMENA

4

DELTA — NIRVANIC CONSCIOUSNESS
- MIRACLE HEALING - CAUSAL CONDITION
- BIOLOGIC & ANATOMIC RESTORATION

7.

You are now going to have revealed to you a visual explanation of the quantum mechanics of our connection to God consciousness via the subconscious mind hologram. First of all, what is a hologram, how does it function, and how is it relevant to our brain?

Hologram

What is a hologram? A hologram is a photographic recording of a light field. Holography is a photographic technique that records the light scattered from a 3D object, splits the beam, projects it onto a photographic plate, and then presents it in a way that appears three-dimensional. In our situation, it is a recording of the physical brain. The following graphic illustrates how a hologram is processed. The interference of the illumination beam and the reference beam create the virtual image (Hologram). The following graph illustrates how and from where the universal frequencies originate and then form a quantum connection to our subconscious mind hologram.

RECORDING A HOLOGRAM

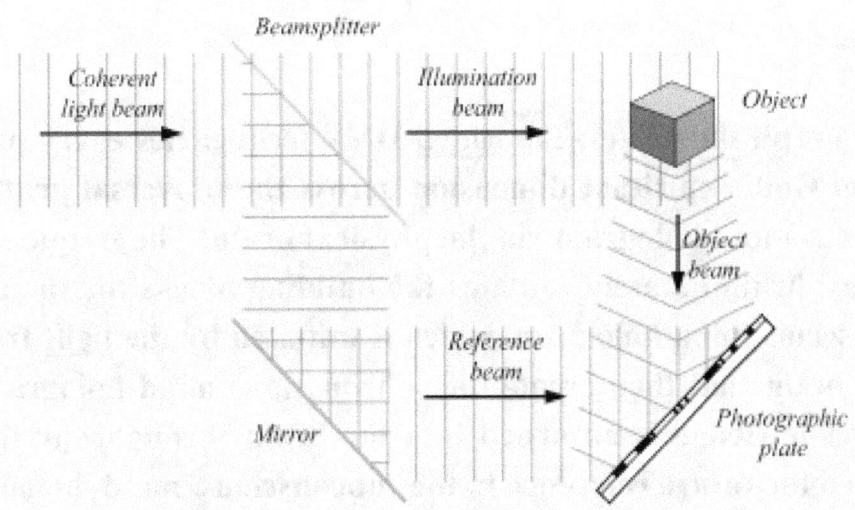

8.

Universal Grid

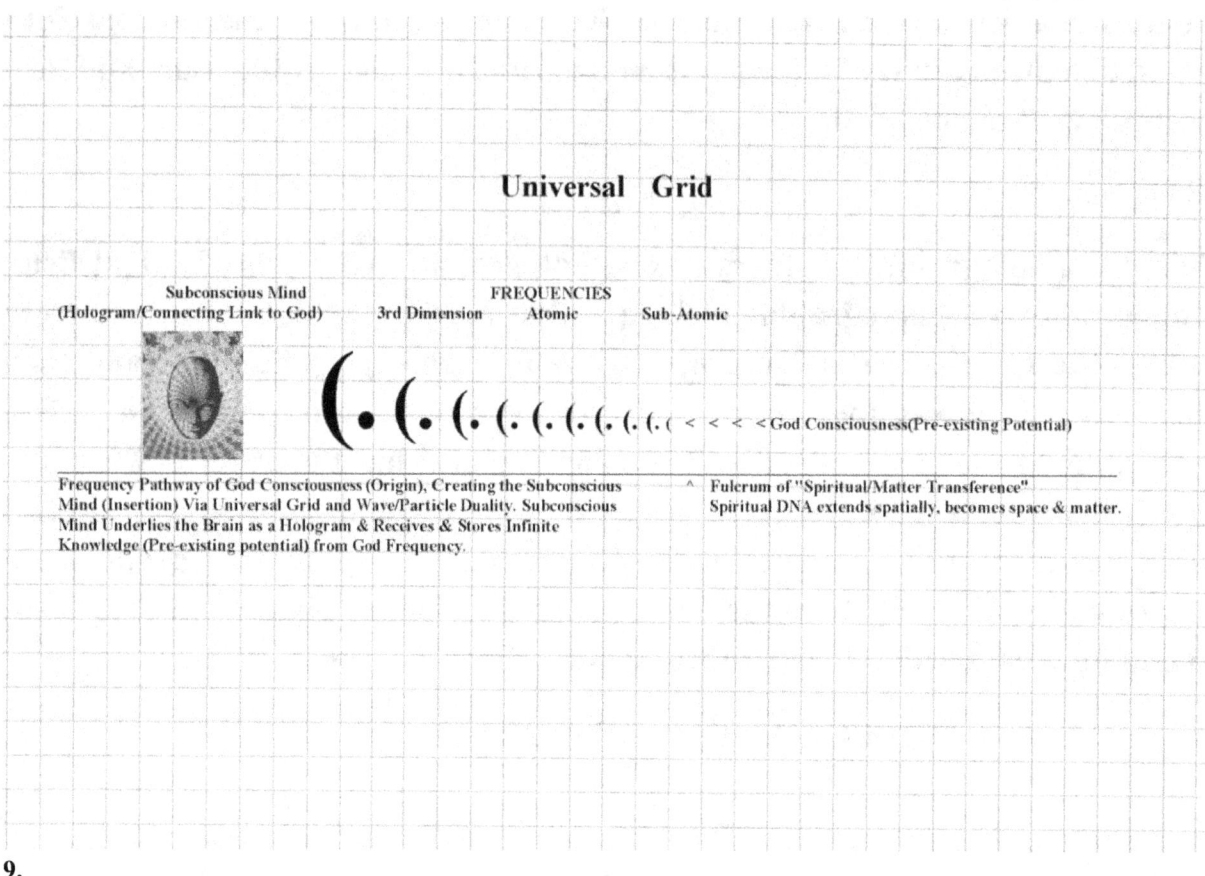

Universal Grid

Subconscious Mind   FREQUENCIES
(Hologram/Connecting Link to God)   3rd Dimension   Atomic   Sub-Atomic

< < < God Consciousness(Pre-existing Potential)

Frequency Pathway of God Consciousness (Origin), Creating the Subconscious   ^   Fulcrum of "Spiritual/Matter Transference"
Mind (Insertion) Via Universal Grid and Wave/Particle Duality. Subconscious   Spiritual DNA extends spatially, becomes space & matter.
Mind Underlies the Brain as a Hologram & Receives & Stores Infinite
Knowledge (Pre-existing potential) from God Frequency.

9.

As the above graph illustrates, the <u>coded</u> DNA frequencies carry preexisting potential from God's spiritual dimension across the universal grid and into the subconscious mind hologram via the physical brain. The frequency <u>coding</u> for the physical brain innately contains the building blocks for the formation of the subconscious mind hologram, which is initiated by the <u>light</u> frequencies beams in the brain that then create the subconscious mind hologram dimension. Further knowledge is absorbed into the brain through our five senses and then stored for future reference in the subconscious mind hologram. The subconscious mind hologram's innate nature is to contain all preexisting knowledge and to translate and supply the brain with the data. It was designed to act as the software and also control all involuntary activity for our bodily functions; it did not have to learn these procedures.

This concludes the "Unification Proclamation: Human Consciousness." It has revealed the interconnectedness of all existing components of knowledge regarding human consciousness, validating a system containing origin, function, and insertion into all universal activity. As a result of understanding this, many theories can now base their validity accordingly.

We have all, to some degree, contemplated God's Master Plan. Maybe God is searching for answers? Maybe this spiritual creation and expansion of the universe is a means by which the Creator can gain knowledge in order to absorb consciousness through our human physical world experience? All of the knowledge that we as humans are experiencing, moment-by-moment, is directly connected to the consciousness of God. All actions that we encounter are registered within ourselves and then made accessible to God's Consciousness through interconnectedness. Perhaps, God established spirit within all physical world forms for the purpose of seeking a new evolution plan that plays humans as "spiritual athletes on the cosmic chessboard." God only knows.

The Secret to Spiritual Healing

I am sure that you have heard of someone who has been stricken with a disease that our present-day medicine has no cure. Then, for some unknown reason, the patient's doctor claims that the disease has completely disappeared. From the knowledge that you have gained thus far, what do you think occurred? Well, with Spiritual DNA extending Itself from the point of Spiritual-Matter Transference, as seen in the Human Consciousness Factor, you can now visualize how the healing power of Spiritual DNA vibrates from Its source, through the sub-atomic dimension, through the atomic dimension, and then into the cellular dimension of our physical body.

With God's spirit extending into everything, the science of spiritual healing commences with meditation/visualization, and strong belief in the healing vibration of delta brain frequency (0-4 cycles/second, Rhythm Frequency Chart). Remember, every portion of the human body has spirit as its underlying foundation, God's Spiritual DNA. The brain vibrates at delta frequency during spiritual healing and communicates with every part of the body by sending messages via motor neurons.

Remember, the brain and its tissue are this outer, physical world's connection to Spiritual DNA, the underlying influence. Also, Spiritual DNA underlies the cellular structure of the diseased tissue. With this understanding, you can, with continuous repetition of meditation and visualization, convert to healing mode.

Take away the inner realm of God's spirit, and the physical universe would collapse and disappear in an instant. God's spirit is immutable. This unchanging power of God maintains all that we know. Ever-pulsating with timely rhythm, the vibration of spiritual consciousness created matter and all that matter encompasses while sustaining creation that was designed to be infinite.

With your mind focused on the visual picture of this pathway, it is clear to understand that the healing force is already contained within our being. It now needs to be activated from within and permitted to begin healing at the cellular level. The healing process can begin with visualization of the diseased area. With eyes closed, visualize Spiritual DNA as the connection vibrating outward from God's source as spirit and into our spirit within. Our spirit within vibrates the healing frequency outward into all three levels of our physical being: sub-atomic, atomic, 3rd dimension, every level of the diseased tissue.

The spiritual realm of the tissue is vibrating from Spiritual DNA. It is Spiritual DNA, because Spiritual DNA created it. When you realize this connection, the power of Spiritual DNA will begin vibrating its healing frequency and become your new default-mode.

In Chapter 10, you will learn of the "New Humans." In particular are the "Children of Aids" who, through the natural process of our DNA evolution, have resistance to all disease known to humankind! More about this phenomenon will be discussed later in the book.

Let's clear the mental fog from our minds, once and for all, that we are the only creations that have the vibrating spirit of God. Spiritual DNA is the makeup of the physical universe. Spirit is not just reserved for we human beings on this planet; It extends Itself spatially, to reside within all life. It created this life, and Its spiritual, governing body is experiencing Its own infinite evolution.

And so it is with this knowledge that we can unplug ourselves from the deception that we are the only creations in the universe that are evolving to fulfill a Divine Plan. As stated earlier, if we were to be the main attraction in this entire creation, God would not have waited 4.5 billion years for Earth to evolve and then have us appear. No, this evolution is a lengthy process, proceeding one moment, one action at a time. Its own evolution has caused It to extend Itself outward (Vector Causation), spatially, as It proceeds on Its quest of a Master Plan.

It is with this profound understanding of who we really are, the Spirit, our

"True Self," that we can perceive life with a different attitude, an attitude that allows us to realize that it is our perception of reality that creates the matrix within our minds and holds us hostage. It is the mind, full of imaginative thoughts of desire that causes us misery.

Now that we realize how God's love vibrates outward through all the dimensions of existence, we now have the moral responsibility to accept that all life begins the moment it materializes into the physical universe and to carry out our own responsible actions of love on a daily basis. Love can be expressed in many ways and to any degree.

Dynamics of Love

Love expressed for the sole purpose of obtaining something in return is not love. Selfish desire is too many times the motive which underlies the projection of love. We see things every day that are desirous to us: a new car, a house, another person, and so on. This is the affliction that brings misery to our lives. It is fine to have materialism in your life; however, to create an obsession for materialism is destructive.

How do we end this dangerous train of thought? First of all, we must lessen our degree of desire for receiving. This is the irony that has plagued the world of human thought - that to receive all that we can possibly accumulate is, "what it's all about." What it's all about is *giving*. This is the key to happiness and the first dynamic of love.

To continuously go out into the world, every day, with the consciousness of a predator seeking to satisfy self-indulgences by any means possible and then justifying those actions for the sake of furthering one's evolution is not love.

When we receive a gift from someone for our birthday or a holiday, every time we see that gift, we send that person a vibration of love in return. This is the power of giving. This is God's love within us being manifested outward to others; this is when all doors shall be open to you.

The first dynamic of developing love consciousness is to give love, unconditionally. We cannot give with the intention of wanting something in return. Many times prayers are offered to God with the intention of receiving something. People pray for all sorts of reasons: cars, homes, love, health, any kind of materialism. This is not love; this is a trade-off. In actuality, this is sinful. We are here to serve God, not the other way around.

We must make giving the primary focus in our hearts. It is only through giving that we show God our true love and respect. God's primary focus is to give us life, give us love, and give us happiness, but all that is given back to God are requests for more receiving. There must be a halt to such a high degree of desiring. The great law of God is, "Give and you shall receive."

The second dynamic of obtaining love consciousness is not to be afraid to love. I have heard many people say that they are afraid to be close and show their love for someone. To do this is to reject God, one of His creations. God's love is expressed in all things. The love vibration is in constant motion. Without God's continuous act of love, the entire universe collapses into nothingness.

When we focus love from the heart, we are extending God's love to the outermost of God's creation, the material world. It is our responsibility to bestow upon all creation the greatest of all actions. Let go of past love failures. Let go of all negative experiences that have caused you to harbor your love. Raise anchor, and let God's "tide of love" carry you to an ocean of harmony and tranquility.

The third and most significant dynamic of a love consciousness is your idealized image of love. Everyone has his or her own image of how love should be. Some look for beauty, some look for personality, some look for companionship, and so on. Whatever your ideal happens to be, the important part is that you express this idealism to its fullest. Carrying God's love to its highest level is the greatest homage that you give to God. No other action that we as human beings participate in is equal to this. We all need improvement in this department.

This is all part of the evolution of our consciousness, the expression through us of God's love. It is our responsibility as adults to pass on to our children this consciousness of love. Each generation will raise the level of degree of love for one another, until high percentiles of love are expressed on a daily basis. As the human race moves in this positive direction, the degree of difficulty to express our love toward one another in life will lessen, and our quest to return to God will strengthen.

This first chapter was intended to introduce you to the rules of the "Universe Game." With the instruction book now in your hands, you will discover how much more simplistic the "game of life" can become. Use this knowledge wisely and repeatedly. There is a whole new life awaiting you.

CHAPTER 2

FUNCTIONS OF THE MIND

> Measure your mind's height by the shade it casts.
>
> - Browning - Paracelsus

Basic Perception

I want to begin this chapter by explaining how the different levels of the mind function. The purpose for this procedure is to familiarize you with the origins of the mind and along with the aforementioned levels of consciousness in Chapter1 regarding spiritual dimension thinking, you may have a better understanding of its function.

As you may know, we humans only use 10 percent of our brain power. From the moment we are born, we are receiving data into our brains from the environment via our five senses. All that we encounter from birth to the present moment forms what we call our personality.

Nature has its way with us much of the time. We have ideas that we create in our minds about something that we want, and right away the mad dash to achieve those desires becomes our sole obsession. It is like nothing else exists but the object we are seeking. From that moment on, all our focus has but one intention - possess. To possess things is fine; however, when the means to achieve creates negative upheavals in our mind and emotions, it is time to grasp onto a different approach.

The Law of Attraction is not an absolute. Place God first. The only control you have is the mindfulness you have over your thoughts, emotions, and actions. You have no control over all the X-factors that are in constant motion, as illustrated previously with "The Bi-Mass Velocity Factor" and "The Equation of Human Events." If it was, every person would have all that he or she desired. It is my responsibility to clear the veil from those who believe that by simply thinking about what you desire will automatically ensure your success. Even though your thoughts may focus upon a desire, your path will encounter a myriad of people who and events that will be in position to thwart your intentions, not necessarily on purpose, but simply because their thoughts and desires do not comply with yours or your timing of the situation. You have to accept the process included with your desire.

What if God says no? There are times when you will have an idea that you know is a winner, but God has other plans. You may think that because you have applied the laws that are supposed to attract your desires, it will manifest. Usually, the opposite occurs; God takes you in an entirely different direction. That is because you are not qualified to receive the blessing yet. The timing for you to be placed into the universal scheme of things has not arrived. A test of faith, perseverence, and patience is now in order. This adversity is all part of your training to prepare you for the next level.

This factor of timing can be elusive. You must allow it to occur without thinking about it. Remember, time does not exist, only momentum. When the momentum changes, timing will favor you; however, be prepared when the momentum changes again. Those that have achieved huge success in their lives experience greater lengths of momentum that favor them.

Success only differs from failure by degree, "favorable momentum" being the key ingredient. You may possess more talent in your field of expertise; however, another person with less talent may possess a longer, favorable, momentum frequency, thus encountering other momentums (Human) that are of a similar mindset and position in life that can produce desirable results for that person.

The Law of Acceptance

The Law of Acceptance is of Divine importance in these situations. When the continuous series of events appear to be placing a stranglehold on your intentions, accept them as part of your training package to success. Not everyone succeeds to the level that they imagine in their mind, therefore, it is essential to accept life as it unfolds. Focus on that which you have control, yourself, and release the energy that you are applying to outside forces that have the potential to produce anything but what you are desiring.

Achievement of goals must be viewed as a journey. Each step along the path of that journey must be nurtured with love and patience, knowing that in your quest to reach a goal, many changes will occur that will affect your mental and emotional states of mind. There is no straight course of action. Just as a highway that connects one city to another, there are many curves, many hills, many steep grades, and many speed changes. So it is with your path from starting point to finish line (From anointment to appointment).

What is important is that we focus on the everyday, little things that God gives us: bringing us safely through the night, giving us the food on the breakfast table, giving us employment to pay the bills, giving us having hot water to take a shower, giving us family members to love and be loved, giving us strength to go to the gym, keeping the old fridge running when it should have broken down long ago.

It is good to set goals, but goals set exactly the way you want them to be and only focusing on that future is to set yourself up for disappointment. The key is to give thanks all day long, to every, little thing manifesting in the here and now. God binds us and keeps us from desires for reasons only God knows, reasons that are meant to prevent us from engaging in situations where we do not belong. The next time you find yourself getting angry over situations that are stalling and preventing you from keeping a timetable of events in your day, think rationally and consider that the interruptions may be preventing you from being in a serious car crash or some other disaster. The events that you are rushing forward to accomplish will still be there when you arrive; they may be <u>more</u> rewarding than you expect.

This is the trust we need to maintain on our path that leads us to our destiny and purpose. Setbacks are not your destiny; they are just interruptions on your way to your celebration. These are the acts of God's laws that were set into position at the time of creation. These laws act out in our lives through the science of mind. We have the power to connect our minds to the universal laws that govern the universe. Let God's universal laws that exist and vibrate within the universal frequencies be your guide in the manner in which you apply the art of mindfulness to your daily activities. Pay attention to the little things; they are the essential parts of the whole. Let God ride you to your final destination.

Control is the greatest test of achievement. Without control, you are like a boat without a rudder, left to be tossed about aimlessly with no direction. Controlling the mind and emotions is a difficult task. The endless bombardment of external forces along our path in life makes the control of emotions an awful encounter. How can we combat this phenomenon called Nature? To understand the full scope of this problem, let us look in more detail to the action of the previously mentioned "Equation of Human Events." Page 19.

"Equation of Human Events"

In the total scheme of things in our daily lives, there exists an equation of multiple proportions. This equation is similar to the possibilities that occur in the popular lottery games, only greater. In our daily lives we have a large number of people that exist throughout the world, over seven billion. When that number is multiplied times the number of thoughts neurologists claim, 70,000/day/person (490 trillion, worldwide), that each of those seven billion people could be thinking each second of the day times the number of possible results that could occur when all these forces are in motion, it is easy to see how anything is possible and how futile it is to place too much emotional emphasis on what happens to us in our quest to achieve our desires. All desire automatically creates with it an opposite force - misery. This is the law. You cannot have one without the other. This is "The Law of Opposites."

This world is a world of opposites, of contrasts, of dual forces constantly struggling with one another. The wise person knows and understands this truth. For this reason, it is logical that we react to Nature's ways with wisdom when encountering situations and circumstances in our lives. Be patient, and know that these negatives are going to happen, but do not get tangled in them. One negative will breed another, until your entire thought process is completely infiltrated with a mental virus. The greater your desire, the greater the virus can become; it is in direct proportion. Every new level has a new devil.

Be thankful for the negatives that occur. They help you learn and gain experience along your path. If everything was unchallenging, achievement would not be worth having. Remain neutral and unbiased with the thoughts that enter your mind. Remember, it is not what happens to you in life; rather, it is how you react to what happens to you that matters. These things are going to happen. When negative reactions in your mind occur, a good way to respond is to treat these negative upheavals as "pop-ups" that you experience on your computer. Create your own "pop-up" blocker; remain neutral and unbiased.

This can be achieved by training yourself to pay attention to your behavior all during the day. When you sense a situation ready to occur that could cause you to react with a negative emotion, remain neutral and unbiased. After you become proficient at controlling your actions, your subconscious mind hologram will take over as your new default-mode and cancel future, negative reaction of your behavior, automatically. It is no wonder the world contains so many diverse people due to the "Equation of Human Events" mentioned previously. Some humans receive more of a certain stimulus from their environment than others. For instance, as a child, one person may have parents who nurture him quite often, resulting in an adult who becomes a loving individual. This continuous repetition of stimulus becomes the dominant force within a person's mindset for years to come. However, this one stimulus does not solely determine the course of action the individual will take. There are a multitude of other stimuli which that same person will interact with during informative child, adolescence, and adult years. Your inner belief system is going to play a major role in your success. It is the person who can perceive reality with the greatest clarity who will potentially proceed to the highest levels of life.

If you are plagued with living a life where you feel that no one likes you, having difficulty associating with others, spending most of your time alone, and you cannot seem to put your finger on what the problem is. This is a time for self-reflection that you must engage in. Apparently, there is a great deal of internal strife that you must confront. When you experience this type of lifestyle, there is surely an aberration in the system in which you have been applying your life. The way in which a person expresses behavior toward others is what is blocking that person's advancement.

Dealing with an immature, inner life, operating from childish behavior is a sure sign of self-victimization. You are an adult; you are too big to be bound by this system that has failed. You have to put the childish system away. You cannot wait for it to go away, you have to <u>put</u> it away. Your functionality has been severely limited; if you don't put it away, it will put you away. It will place you on the "sideline" of life; you could be there permanently.

You may realize that you possess more talent and potential than others; you know that you should be leading and not following others that you associate with. This is the time when you have to lose your fear of "killing" your old self. You have to first, identify the problem and then remain neutral and unbiased toward the old thoughts and emotions that have led you to resistance-attraction and self-sabotage. It is time to cut off the head of the monster.

No one can prevent all the events that cross our paths from affecting us to some degree. After all, this continuous experience of the path of life is what makes us who we are, is what gives us our many facets of behavior during different situations. Everyone has more than one side to his or her personality. A woman may demonstrate motherly qualities to her children, and then, an hour later, she is in charge of a group of business partners. We "wear many hats" in our lives.

Multiple Realities

During this so-called personality development, Nature has its role in our evolution. Nature being so unpredictable a force can bring a host of merciless events to us as participants in this physical world. How are we going to know what will happen from one moment to the next? Which part of our personality will come to the forefront and act out our role? What kind of world will you create for you? The possibilities are endless.

Forget about parallel universes and the claim that there is a multiplicity of you, each living a different existence elsewhere. Besides, did these multiple universes always exist, or did our creative minds conjure them up? This is a ludicrous theory from the onset. With seven billion people on this planet and growing, the number of universes that would have to exist to accommodate all the variations of situations would be astronomical. A theory such as this is devised by someone who is dissatisfied with his or her life and wishes to believe that another existence of them is experiencing something better.

The only multiplicity of realities that exists is right before you, in the present. Take a look around you right now. How many different physical forms do you observe at this very moment, at every moment of your existence? This constitutes the potential of multiple realities. You have been given the freedom to choose which of the influences before you at any given moment will be your path of reality based on your internal, subjective observation. This is your belief system as a result of your brain conditioning.

You may react to a situation with an uncontrollable display of negative emotion, causing you to place yourself on a path of misery. Your thoughts may continue, causing you to experience more anxiety and anger. Depending upon how much control you have over these upheavals, your condition may go on for long periods of time. Each time you react to life's events in this manner, you are re-enforcing the stronghold that events have over you. You create a world of bondage in which your mind knows no other direction to take.

Reincarnation Put to Rest

As result of understanding, "Unification Proclamation: Human Consciousness" (Chapter 1), reincarnation can now be viewed in a different light. With the knowledge of how Spiritual DNA extends itself through all creation as a result of the laws of physics that Spiritual DNA created at the onset of the physical universe, it is clear to evaluate that all male sperm created for the purpose of extending life will immediately possess spiritual content as its core makeup, as does the egg from the female.

This also clears up the debate concerning when life begins. It begins the moment Spiritual DNA creates it through its own spiritual frequency vibration. As the human form progresses and develops in size, all of the physical attributes of that human being will also contain the spiritual attributes proportionately, as a result of the vibrating wave-particle frequency. At the termination of the physical body, there is no next, physical lifetime to experience, no next, physical body to enter. Each new body has its own spirit automatically created as its core by Spiritual DNA frequency. In essence, our belief system has been in reverse. We are not a human form created at birth, only to have a spirit inhabit our body. In actuality, the frequency of Spiritual DNA vibrated the origin of our human form as sperm. Remember, in Chapter 1, Spiritual DNA extended itself and created all levels of the universe? This includes all reproductive systems; all physical world vibration contains Spiritual DNA as its base.

Before the physical universe became what we witness it as, it was a pre-existing, spiritual potential. The entire universe was spirit before it vibrated to physical matter, just as our body began as spirit and vibrated outward to the physical. In other words, there are no past lives or future lives in other bodies. Each body has as its underlying vibration its own spirit. This holds true whether you are a creationist or an evolutionist. In the creationist belief system, how did the first humans have a former, human life enter them if there was no humanity previously? It surely did not come from a former body. This holds true no matter where you go in the universe where there may be alien life forms. The first life forms in the universe had nothing previous to it. In the evolutionist belief system, evolving species would still have an individual spirit at its time of conception and proceed to higher levels through DNA mutations. The potential to become what we become is already <u>encoded</u> in our spirit through our attachment with Spiritual DNA frequency.

What happens to the spiritual core of the human that passes? It remains a part of the force of God within the spiritual realm, the white light seen in "near-death," experiences, to remain for eternity as it always has, as infinite, spiritual consciousness. Our spirit is connected to Spiritual DNA frequency. It is not responsible for the actions that the physical body acted out through decision-making as a result of the physical body's (Brain, emotional conditioning, behavior) perception of reality. Our spirit does not act itself out through our five senses as expression. The soul performs its expression through the brain/emotions/behavior.

With the brain acting as the physical influence of the body and the mind and emotions acting as the mental plane influence, retribution for all actions must be accounted for upon passing of the physical body. It is the mind and emotions that comprise the soul of a human being, environmental influence and <u>ancestral coding</u> in our DNA being the key factors. Soul expression in life is a matter of degree. Some express their mind/emotions with deeper content than others, thus more soul is the result. The soul and the spirit are of the same Spiritual DNA; however, spirit is the essence of our "being." The vibration frequency of the soul (The brain, mind, emotions, and ancestral DNA coding) is the result of the inner spirit (Origin) vibrating outward spatially but is influenced by free will.

Every action is recorded in the grid (See Akashic Records, Chapter 8). If your belief system condones an existence of Heaven and Hell, then atonement of souls would vibrate at a frequency within the grid. Heaven and Hell would be of the same polarity, only differing in the degree of atonement. Reward and punishment would be of a mental/emotional (Consciousness) nature, to last for eternity. Reward would be eternal existence in light frequency (Heaven); atonement for sins would be assessed in the form of intense, heavy, density frequency (Hell), possibly in a heavy density 4th dimension. Page 13 Frequencies Permeating the Physical Universe

CHAPTER 3

BEGINNING YOUR ASCENSION

The mark of rank in nature is capacity for pain.
And the anguish of the singer marks the sweetness of the strain.

- Sarah Williams

First Steps

In this chapter, it is noteworthy to begin aiding you with the fundamental knowledge and understanding necessary to begin your ascension. I will begin by introducing you to the most basic and fundamental principle that I have taught during my many years as a metaphysics, research scientist - the power of breathing.

Healing negative thoughts and emotions is not an easy task. Remember, you have been involved in this train of thought and action for many years. However, you may begin the transformation from this state of being, immediately, with one single, miraculous movement - breathing. But, you say that you are always breathing. How can this be anything new? You are right; it is not anything new. What is new will be the way you perform the act of breathing.

Have you ever watched a baby standing with belly protruding, the belly moving inward and outward? Babies have the purest breath action. You may ask how a baby can know anything about its breathing. That is the point. It does not know. A baby does not have any reason to engage in improper breathing habits; it functions with pure, innate action produced by the subconscious mind hologram that controls all involuntary body activity. A baby does not think about how it is going to pay the rent, where its next meal is coming from, why all the signal lights are turning red when it needs to get somewhere on time and soon. It is only when we become adults that we begin the journey into the realm of misery created by our thoughts and emotional upheavals, and that allows our timing in life to turn our path into a series of misfortune and bad luck.

Timing

We hear that word all the time. Usually the people who say they have good timing are those who are not in any hurry to get somewhere. Even if you are hurrying around to accomplish something and do succeed, something else will eventually slow you up. This is the way things are. The laws of balance are in constant motion; they work for everyone, some more than others, but we all get "balanced out" sooner or later.

All of you probably have experienced what I call the "Red Light Syndrome." You are traveling in your car along a street in town and suddenly, another car swiftly passes you by, only to have to come to a stop at the next corner. A modern version of Aesop's Fables, "The tortoise and the hare." Hurrying to achieve desires is senseless. You cannot know what is ahead in the future. Only your mind creates what you would like to happen, and when it does not happen in the time-frame you demand, negative emotion is the result.

There is only one way to create good timing and change the endless search for that great stroke of luck in your life that you think is going to bring you the true happiness you have imagined in your mind. You must learn the proper breathing technique, and learn to relax and enjoy God's creation as it already exists.

Remember the baby and its breathing posture? As we grow, we are taught that our breathing involves the upper torso, particularly the chest area, because that is where our lungs are situated. We practice this method of breathing for years, developing our chest muscles to expand and contract with each inhalation and exhalation. When we engage in situations in life that involve tension, we have the tendency to tighten up our upper bodies and even stop breathing. This lack of attention to our most vital of all biological actions is to ask for trouble.

Not only does this behavior create a loss of oxygen to the brain and all parts of the body, the tension that develops and continues each time a similar situation occurs throughout our lives will cause disease to erupt. This area of infestation will be different in each of us. It is dependent upon which part of our immune system is the weakest. This is how cancer and serious heart conditions develop in humans.

Now, you are probably asking yourselves, "You mean that something we have been doing all this time is wrong?" Well, it is not wrong, it is only dislocated. What we want to achieve is the same breath; only, we want to lower the action a bit. This may seem a little discomforting at first, but with a little attention, the proper movement will be achieved.

If you learn only this one technique, your entire life will manifest into a miraculous series of events that will place you in harmony with everything in Nature. No longer will you be fighting Nature; you will find yourself flowing with the tides of natural phenomenon. The breath of life calms your biological system to comply with the laws of God. Each individual will find his and her natural breathing rhythm.

The breathing cycle is different for each person. You must practice this form and find what is natural for you. There is no fast pace, no slow pace, only somewhere between for you to discover. Begin by standing with a straight posture. Do not make yourself feel tense. When you begin inhaling oxygen, allow your belly to expand, not your chest. After the belly expands, slightly expand your chest, and then move your shoulders back even more slightly.

The reverse is now the movement for exhalation. Allow your breath to come out now, lowering your shoulders, then your chest, and finally your belly, paying close attention to these movements as you practice. After repeated sessions, the automatic mind (Subconscious mind hologram) that controls all your vital functions will take over, and the thought of having to remind yourself to breath properly will no longer be necessary. It will be as second nature to you as your old habit of breathing was.

As time passes and you begin to feel comfortable with your new cycle of breathing, a new series of possibilities will occur in your life as a result of the calming effect proper breathing brings to your thoughts and emotions. Calming the mind and emotions creates good timing and good luck in your life, because you have found your own level of breathing that deletes old influences in your mind of how you should live your life. This occurs quite miraculously. No longer will you be trying to force things to happen for the sake of accomplishing desires according to your time schedule. You will simply allow the events to happen in the sequence and timing that Nature is unfolding them. Instant gratification will no longer be a part of your imagined reality, as a result of society's influences. You will simply accept the way things are in the universe of God.

We never know what is really good for us at any particular time in our lives. We only imagine what we think is right for us. We have been creating this imaginary playground all our lives. Our finite minds cannot possibly know what God knows. We are in the positions in life, because we belong there at this particular point in time in our personal development and the evolution of the universe. Nothing that we imagine in our selfish, little quests for desire is going to change the sequence of events from happening the way they are going to happen. So, stop being the fly that bangs itself into the window. Remember your breathing and flow with the natural rhythms of God's breath of love.

Gratitude

Another realization that one must practice on a moment-to-moment basis is giving thanks for the privilege to be an active participant in this physical world. Now that you realize how your thoughts are constantly affecting the quality of your happiness and the only way to achieve true happiness is to pay attention to the moment before you; be so enlightened to the very beginning of your existence.

For as long as you can remember, you have been living your life, day-by-day, with the attitude that you are some helpless victim of some universal plot to keep you in the prison of misery. The truth is there is a universal plot, and it was designed to unfold in a way that allows you to eventually come to terms with the entire universal scheme of events. Nothing is out of position in the universe at any particular time. Everything that occurs in life is in direct proportion to all that has occurred before. Think for a moment all the seconds that has passed in all of time before this very second and all the seconds that will come to pass for all eternity. I realize that this type of thought can be mind-boggling; however, it is necessary to help you put your own life into proper perspective.

If we stop for a moment and each of you thinks back to the day of your birth, you can then begin to realize the enormous chain of events that was necessary to allow your present existence to manifest. In the process of conception of birth, hundreds of millions of sperm are released to engage at the precise moment of conception with the egg. The probability of your form, out of hundreds of millions of others, attaching itself and becoming a living organism is so great it make the money lottery we play on a weekly basis seem insignificant in comparison. And you always considered yourself such an unlucky person? Do you now realize how lucky you really are? Imagine all those unfortunate forms that could have been instead of you, and why, because it is all part of a universal plot? Or, maybe it's a universal play?

Everything that has gone before is totally necessary for what is happening right now. Do not curse the moment for what it is showing you. Bless it, and know that it is of great importance in order for the next moment to unfold. When you show this kind of respect for the creations of God, the entire universe shifts, even though ever-so-slightly; it shifts to make way for you to receive God's return. This is the way it is.

This great realization of the miracle of our own lives is the key to the world of human beings coming together, an enlightenment revealing that each of us who exists at this particular time-period in the history of the universe is a part of everyone else. We have the opportunity to celebrate the life of all on this planet. If we are to proceed as a human race into the future, we are going to have to express and apply this miracle of life that God has given us all. This is not going to happen overnight. This is a tremendous task that must be the responsibility of everyone. If you want a consciousness revolution, the only solution – EVOLVE.

We are presently living in an age of communication. Great! So, let us begin communicating. Start carrying the message of this evolved consciousness to your own mind and actions first. When each person takes a look in the mirror and reflects this new attitude to the spirit within, an action of goodness will begin to form that will gain momentum and spread throughout all humanity.

With the means of communication we have at our disposal throughout society (Cell phones, internet), we can accelerate this process of higher consciousness, exponentially. We are the humans who, at this point, are at the threshold of a breakthrough in the way we prepare future generations to carry on the quest to live with the truth that we are all miracles of God.

Play Reverse Psychology with Nature

I have developed a simple method that will allow you to put aside the eruption of negative emotion that usually accompanies desire of physical world phenomena. When Nature attempts to block your entry to the door of success, turn your attention away from that which you are having difficulty achieving, and place your energy on some other activity. You are trying too hard.

There is a law called "The Law of Reverse Effect," which occurs when people over-kill their efforts while striving for accomplishment. Many athletes are guilty of this forced effort. They allow their mind to interfere instead of letting the subconscious mind hologram control their functions. The harder you try, the more this law goes into effect. Just say to yourself, "I am not going to hold on to this over-indulgence for this want that I have. It will happen when the forces are right to bring it to me."

When you set this mental attitude into action, you are telling Nature that your mental imprisonment for this particular desire is unlocked, that you are releasing yourself from it. As soon as you take your focus away from your desire and place your attention on another desire, Nature releases its hold on your original desire and begins its holding game on your new desire, at the same time allowing your previous desire to be free to unfold and manifest into your life. This action relates to the previously mentioned "Bi-Mass/Velocity Factor." You must realize that all action beyond you is occurring simultaneously to benefit you and send that desired momentum in your direction.

I know, you have been taught to persevere until your goal is accomplished. Perseverance is a virtue, I agree, and it brings misery along with it. What I am saying here is when persevering toward your goals, be willing to release yourself from your entanglement now and then. Remember, the only way to defeat Nature is to not desire anything. However, when you need to indulge in desire, do not be overwhelmed by it to the point where you are consumed with stress when your desire does not appear as you have scheduled. This is the underlying cause for our indulgence in instant gratification. Society has become so consumed with instant or near-instant gratification in their lives that our anger over not having a clear path to our destination on the "roadway of life" leads to irrational behavior. My God, lighten up!

You do not know what lies around the bend. Did you ever think that the few red lights you had to stop for might be delaying you so that you will cross paths with someone who will bring you good fortune? Think about these things the next time you become angry when traffic slows down and you have to wait a few minutes, or for any inconvenience. These occurrences happen for a reason, at the precise moment when they are supposed to happen. They cannot unfold in any other way or at any other time.

Nature is not going to change for you, for me, for anyone. Understand the "nature of Nature;" it is as it is, unfolding with undaunted precision and perfection. It is only with this realization that we can live free, moment-to-moment, and enjoy all that life gives us.

"If You Yield, You Will Overcome."

The most difficult of all human behaviors is the act of yielding. Again, people are in a hurry to achieve this and that, unaware of a universal truth that acts in direct contradiction to what we are led to believe.

One evening in 1972, I was watching an episode of the then popular "Kung Fu" series. David Carradine's (Kill Bill) Shaolin monk character, Kwai Chang Caine, was revealing to another person the wisdom of yielding in order to achieve his goal. "If you yield, you will overcome."10. This very simple yet profound truth is key in accomplishing victory in your life. Think of the numerous times during your daily lives when you are challenged by the forces in your environment. At home, at work, on the road, everywhere you go presents potential confrontation to some degree. Again, it is not what happens to you that matters; it is how you react to what happens to you that is important. This means mental and emotional control. The phenomenon of this physical world is never going to change. Pesky, little irritations are going to continuously occur wherever you go. It is your responsibility to bring about the change.

In every interaction between you and life, there is that which is perceived and that which is the perceiver. You are obviously the perceiver, perceiving all that you encounter each moment of your life. When engaging in these situations and the red flags begin to appear and preventing you from accomplishing your goals, you must react with the proper attitude. Do not perceive what is occurring as something containing a negative nature. The sequence of events before you can be no other way because of everything that has unfolded before. The series of circumstances are perfect as they are. It is now up to you to adjust your point of view concerning these matters.

Yield to that which you perceive, allowing it to take place without expressing negative, emotional reaction. The moment you, the perceiver, choose to accept this, the entire circumstance before you will take on a new perspective. Remember, objects in motion tend to stay in motion and along with this truth, tend to change. Your new perspective will now be treated to a sequence of a miraculous coming together of the elements involved in bringing to fruition your intended goals.

Yielding on the Freeway

The action that occurs while driving our cars is probably the best analogy I can use to demonstrate this phenomenon. It does not take many annoying actions by other drivers to erupt a negative, emotional reaction from us. When we lose control in this manner, the phenomenon of action/reaction automatically continues, along with other unsuspected circumstances.

One after another they persist, until you fill yourself with rage. Begin by sending your mindset to the other end of the thought spectrum. If you are driving on the freeway and another driver darts in front of you, remain calm, knowing that this situation could only occur this way due to everything that has gone before it. By yielding to this situation in this manner, the phenomenon of action/reaction will begin to unfold in your favor. Any number of possibilities can immediately take place. Remember, everything is constantly changing; all the other cars around you have the potential to change in your favor and create a situation beneficial to you. You will be amazed how the entire scenario before you will open up and allow you to be in a better position than previously.

Another yielding scenario that you can practice while driving is with traffic lights. Many times you may be in a hurry to reach your destination, only to be denied at every intersection by red lights. By training your mind to accept these unwanted repetitions of traffic signals, the action/reaction phenomenon will be activated in your favor. In accordance with the universal "Law of Balance," it will counteract with a series of green traffic signals ahead, which will give you the necessary make-up time needed to reach your destination.

I encounter situations, every day, where someone will yield. My favorite was a situation in a supermarket. When I went to the check-out line, three people were ahead of me. The lady in front of me had a shopping basket full of groceries. When she saw that I only had two items in my hand, she gestured for me to go ahead of her. I thanked her. No sooner than she acted with a consciousness of yielding, a check-out person arrived on duty and asked the next person in line to come into her aisle. It was the lady who yielded to me.

She yielded and then overcame the long line. She even finished before I checked out! I had to approach her afterward and tell her that she had participated in an ancient philosophy. She said that she would remember to repeat it in future encounters.

When this knowledge is realized after engaging in this small action in your life, the same principle can be applied to all your life interactions. Think of how many times during the course of your everyday living situations that you will be able to change seemingly stressful events into enjoyable encounters. When harmful, undesirable thoughts appear at the forefront of your consciousness, send these unwanted frequencies to the opposite end of the polarity pole. Everything in life has its opposite; however, the opposites co-exist in the same polarity. They only differ in degree of frequency. The following picture image ex-plains this phenomenon.

Degrees of Separation

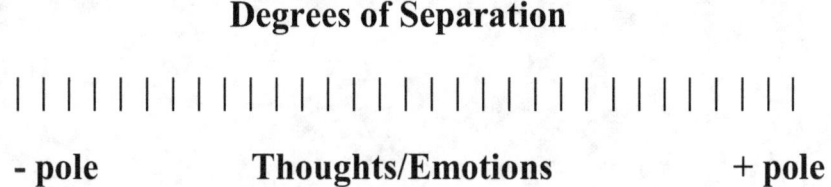

- pole   **Thoughts/Emotions**   + pole

Thoughts and emotions surface continuously from influences in our minds. They will move like a sliding scale from one pole toward the other by degrees, depending upon the intensity of the circumstance you experience. The "Pendulum Effect," as described in Chapter 2, pertaining to the conscious/-

subconscious mind hologram action has a role here also. Positive and negative influences swing back and forth. When an unwanted, negative vibration of thought or emotion moves in the direction of your positive consciousness, remain neutral and unbiased. The influence will return to its far end of the polarity pole. This repetition will allow the subconscious mind hologram to keep the undesirables at a distance. This strategy of living is the start of a grand epiphany for you.

CHAPTER 4

ASCENDING TO THE NEXT LEVEL

Wake up everybody, no more sleepin' in bed.
No more backward thinking,' time for thinkin' ahead.
The world has changed so very much, from what it used to be.
There's so much hatred, war, and poverty.

-      Harold Melvin and the Blue Notes

Early Stages

As you are beginning to discover, the limited use of our brain to this point in our evolution has a tendency to place our observation of reality on a low plateau. For this reason, if for no other, we must train ourselves to keep an open mind to all observations and experiences by persons who have had encounters that have not yet been placed under the scientific microscope. We are constantly plagued by the perceptions we make of our external environment, which keeps us from perceiving the truth in all things.

The closest we have ever been to the truth was when we were children. Our spirit then was strongly connected to the spirit of God, but as we aged and became influenced by brain conditioning, the competitive ego began performing its separatist power over us, blinding us from the Light and casting a mist over our "True Self" – the spirit. For this very reason, Western psychology is invalid, for it only treats those aspects of human beings pertaining to the mind/emotions and the influences they have incurred from their perception of Nature. The time will come when Western psychologists accept the fact that, their treatment must extend into a deeper realm of the human element, the spirit. Then and only then will the origins of all mental illness be accessible and open to healing.

It is necessary to realize that, our trek through the powers of the mind is only in the cradle stages of development. Just as all great thinkers and inventors were scorned initially for their beliefs, so it is today that ignorance prevails in much of our world.

Ascension: By Lisa Renee 9.

"Let us review the meaning of ascension and why we as humans will experience physical, emotional, and mental symptoms and changes. Recently, we have been in a transition phase of expanding into a realm of higher consciousness as multi-dimensional human beings. During this phase of the Ascension cycle, human beings will experience many varied physical symptoms as the cells in our body shift into higher frequency patterns.

Ascension is about bringing the layers of light, the force existing within the levels of our spiritual bodies (Our multi-dimensional God-selves), by descending these layers into matter and anchoring them into the physical plane of reality. Ascension is a shift in energy frequency patterns held in a dimensional space which when absorbed and activated into the layers of the planetary and human bio-energetic field, activates its DNA template instruction set. This catalyzes a chain of events that creates a complete transformation and transmutation of various patterns and programs held in the energy templates of the human soul's journey within a cycle of evolutionary time. When activated, these patterns begin to shift, re-emerge, and then clear from the layers of experience coded into every cell and memory pattern held as an energy vibration within the body.

The ascension process is about moving our consciousness from one reality to another and the awareness of possible, multiple realities existing simultaneously. Since a "reality" is a dimension, what we are undertaking is in essence a complete dimensional shift. A dimensional reality is held in place by a complex layer of coded energy grids to create the illusion of time and space for the consciousness to perceive and participate in its own experience within the broadband widths of that particular range of frequency.

The dimensional grid shifts its frequency, and its magnetic attributes change; all things existing within that broadband reality will also shift and change in a myriad of ways. The natural laws governing time and space as we know them will change. This also means the perception of our spatial awareness; our relationship to time and space will also change rather dramatically.

As we go through this shift, we must prepare and adjust our way of thinking and our entire being to that which is in alignment with our "soul" purpose and true, divine essence. Surrender and acceptance are two main characteristics we will need to allow permeating through us to facilitate an easier time. In order to shift the old behaviors and thought patterns not serving the "soul" purpose, we will become aware of the beliefs and behaviors that exist as imbalances within us, and we must take the appropriate steps to integrate or clear them. This process will bring our deepest fears, beliefs of limitation, and old pain patterns to the surface through events that trigger them into awareness, so that we can consciously acknowledge, resolve, and heal them.

Some of these issues are ancestral or inherited in their nature and may feel rather odd yet familiar when they are brought to your awareness. It is important to remember that we are not only clearing our individual mind grid but also the karmic mind implications and collective unconsciousness implications on this planet from the last evolutionary cycle. This way you can reframe the circumstance in your mind to be that which is impersonal and detached from identifying with the specific issue that has risen to the surface to be cleared within your body.

Appreciate and acknowledge your "light being" for accepting the assignment for this clearing rather than claiming it as belonging to you. Doing this repeatedly, we clear our emotional and energy bodies of the long held traumas, beliefs, and fears that have impeded our ability to experience our "true being" and bring more joy into our lives.

Once the emotional body is resolved and cleared of an old pain or trauma, the physical body then is enabled to clear its equivalent of that block. This happens simultaneously within all the layers of the bio-energy field when each healing is cleared. In effect, it is a full clearing and release of these painful events held as patterns in the body from the soul's record and cellular memory. And that is pretty awe-inspiring when you realize how this will impact the human experience as we move forward.

Eventually, as these painful patterns are removed from the soul record, the clearing recodes our DNA, so it can hold more light, thus sustaining our new frequency and a state of increasingly higher consciousness and multi-dimensionality. As we hold this greater light and advanced genetic package in our bio-energy field, it enables others, through the principle of resonance, to vibrate at the increased frequency rate and activate their own inner potential and divine inheritance towards Ascension.

Unconditional clearing is the main process we will experience and be acutely aware of as we shift through the Ascension. Since emotional pain and imbalances are responsible for our states of "disease," many of us will experience illnesses or physical releases of pain clearing from our bodies. Some of this clearing is for your karmic, genetic, or ancestral soul lineage. Some of us, as light workers, have chosen to clear states of imbalance for the collective or for a specific group and will feel this movement of energy transmute through us as our energy being is acting as a "cosmic filtration system" for the greater whole.

The goal of this series is to inform you of these dynamics that we are currently experiencing, as the energies accelerate to prepare you even further by clearly discussing the various, possible, physical symptoms from a position of having a "neutral" association. Utilizing what works for you personally by cultivating intuitive discernment with an informed awareness is the intended support of this discussion. This is not to alarm you but is used as a tool to keep you informed in order to be self-accountable and self-sovereign in all your upcoming choices.

We will not be given anything we cannot handle. Much of this can be avoided by being self-aware and taking the time to heal through emotional clearing and listening to your body when it requests other healing techniques. Spirit has given us many tools for our Ascension toolkit, to assist and support our evolution in the easiest possible way. Ask for assistance, and it will be given!

Transmutation Systems

Owing to some of the extreme transmutation systems that are rampant at this time, I want to be specific, and this may be more lengthy than required. I get clients, family, and friends that feel that something is very wrong in what is happening to them because of the extreme transmutation and transmutation systems they are feeling. Many are vague to describe and inexpressible in linear thought and language. When they go to doctors, they are being told to take drugs, or they are being instructed to seek professional, psychological help. Of course, I am not professing to be giving any medical advice. It may be beneficial for some of us to seek out medical or psychological treatment during our process. We must follow our own, inner guidance on how to proceed with what is occurring within our physical and energy body systems at this unprecedented time of acceleration.

Certain symptoms are now becoming more apparent, and when we know this is a "process" and that many others are feeling the same things, certain serenity can be brought into the mind. Many can rest in the knowledge that they are being transformed in ways that may feel extremely hard on the physical body. Some of the symptoms reported are:

- Cranial pressure and headaches

- Extreme fatigue

- Heating up of the physical body

- Nausea

- Dizziness

- Forgetfulness

- Irritability

- Sleeplessness

- Joint pain

- Body aches

- Muscle cramps in legs, calves and shoulders

- Flu-like symptoms

- Kundalini experiences

- Diarrhea

- **Feeling out of sorts**

- **Muscle pains**

- **Skin rashes**

- **Tingling in body parts**

- **Having specific awareness of an internal organ or body part.**

- **Diminution of spatial awareness**

- **Clumsiness**

- **Feverish feeling**

- **Feelings of being there and not being here**

- **Loss of visual acuity**

- **Memory loss**

- **Changes with body and head hair**

- **Lack of ability to concentrate**

- **Feelings of standing up or moving too quickly that create vertigo.**

- **The feeling that you cannot accomplish anything; not enough time.**

- **Anxiety attacks that happen suddenly and disappear just as suddenly.**

Also, our bodies often experience a rush of heat and energy bursts that are not comfortable. Our bodies are being shifted in thermodynamic ways and with so much light entering the fields of the body, some of the symptoms we experience are actually preventing our physical vehicles from bursting with intensity of light. We may experience a "triad" sleep pattern (Waking up every 3 hours) or interrupted sleep as we are being recalibrated and worked on energetically at night. This adds to our being restless and tired in the morning.

Remember to be gentle with your hearts and entire being, and only allow others within your boundaries of inner sanctum that treat you the same. This can be an extremely vulnerable time, emotionally. We need to release the fear of asking for help and investing in ourselves to feel balanced and healthy. Massage, Bodywork, Energy Work, Spiritual Counsel, and just plain laughter and FUN are extremely important during this time. Allow yourself!

If I could impart anything to our readers, it would be to emphasize how humans are so deeply loved and appreciated by Spirit, the Hierarchies, the Planet, from all those of whom we serve as One. The impact of our collective work during this evolutionary time cycle is beyond comprehension! Thank you with deep gratitude for all of your amazing work! Stay in the luminosity of your heart and soul path! We are here as One!"

Effects of Grids

Shakura Rei explains the grids and their effects on human beings regarding geopathic stress. "'Geopathic stress' may be defined as developing bodily imbalance or illness due to lines of harmful energies which radiate from the Earth. Earth acupuncture is the process used to neutralize these fields. Geopathic stress has been found to be the common factor in many serious and minor illnesses and psychological conditions, especially those conditions in which the immune system is severely compromised.

"Cardiovascular deficiency, attention deficit disorder, immune deficiency disorders, chronic fatigue, and cancer are samples of chronic geopathic stress influence. Some lesser effects of influence are chronic body pains, headaches, and sudden signs of physical aging, irritability, and restless sleep. It is also a common factor in cases of infertility and miscarriages, learning difficulties, behavioral problems, and neurological disabilities in children.

"When treating patients who continue to be affected by Earth radiation during periods of sleep or work, response to treatment tends to be slow or uneventful; however, when their home or work place is neutralized, the geopathic stress conditions resolve themselves and the body begins to heal.

"There are four lines of geopathic energy which can be detected and neutralized. The Hartmann Grids and Geopathic Zones adversely affect the cellular structure of living organisms. Interference Zones and Personal Zones affect one's emotions and objectives. Personal Zones may also affect one's physical health."

1. What are geopathic zones?

"Geopathic zones are places on the surface of the Earth which can cause serious health problems for people who stay within them for a long time. The name geopathic comes from two Greek words, geo-earth and pathos-suffering.

"Since the ancient times every nation had their own perception of "dead" places, where trees and grass do not grow, where people and animals get sick, where buildings fall apart. For many centuries, people were thoroughly careful when choosing places for their dwellings.

"The Chinese system of feng shui has been around since the times unmemorable. In accordance with this system, no one would start building anything until a dowser made sure that the place was not under the influence of some "underground demons." Ancient Roman architect and builder Vitruvius paid special attention to the correct selection of a place for building in his works.

"Similar principles are also mentioned in the works of Hippocrates and Avicenna. In Russia in the 18th and 19th century, the selection of a place for a house was entrusted to a dowser, and the decision about construction was made by the Czar's decree. Over time, these important principles were forgotten, causing problems and neurological disabilities in children. Houses were built in convenient places, which since the old times were considered unsuitable for a habitat."[10].

2. How does a geopathic zone affect a human organism?

"One of the first researchers interested in connection between geopathic zones and origins of some serious illnesses was Gustav von Pohl. In 1929, while observing 10,000 patients, he came to a conclusion that the one common thing for 58 people who died from cancer was that each of them had a sleeping place (bedrooms) exactly within a geopathic zone. The scientist thoroughly described his researched results in his book, "Earth Rays as Pathogens."

"In 1950, Manfred Curry, MD also came to the conclusion about an important role of a special earth grid (Later named the Curry grid in his honor) in development of cancer. A fundamental book by Dr. Ernst Hartmann called, "Illnesses as a Problem of Location," which was published in Germany in 1960, summarized many years of the author's research of geopathic zone's influence on human health. In this book, Hartmann was the first to introduce principles of projecting and constructing buildings with consideration for geopathic zones.

"For 14 years, Austrian scientist K. Bahler observed 11,000 people, including 6,500 adults, 3,000 teenagers, and 1,500 infants. Obtained results showed that cancer, neuropsychological, and various chronic illnesses of kids and adults were caused by their bedrooms located within geopathic zones.

"Since the early 1980s, scientists from Austria, Germany, USA, Switzerland, England, Canada, and France have been studying the problem of geopathic zones. Many years of large-scale observations and research showed that:

• Depending on the longevity, the nature, and the location of a human being within a geopathic zone, various illnesses set in, affecting body organs and disrupting their functions.

• The most common illnesses are of oncological, cardiovascular, neuro-psychological nature, as well as disruptions of the motor functions of the body. If the entire, human body is within a geopathic zone, it affects the entire system, including joints, circulation of blood to the brain, and causes sclerosis, sores which do not heal.

• There exists a certain period of time after which a human organism becomes misbalanced, leading to further development of pathological conditions and eventually death.

• All people within a geopathic zone have one thing in common and that is their complete insensitivity to any kind of treatment. It is practically impossible to heal a person within a geopathic zone."

3. How to detect a geopathic zone.

"It is possible to approximately locate a geopathic zone by watching animal behavior. For example, dogs avoid such places, but cats prefer to rest in them. Horses, cows, and pigs also avoid geopathic zones.

"Among the trees: apple, lime, beech, and lilac usually grow better outside of the geopathic zones, but plum, cherry, oak, ash, spruce, peach, and mistletoe tend to be closer to geopathic zones because of underground water.

"In Nature, trees with swellings, anthills, as well as trees hit by lightning bolts are the signs of a geopathic zone presence. Flowers, such as azalea and cacti grow better in safe places, while asparagus and geranium prefer geopathic zones.

"Reactions of a human organism to a geopathic zone can vary. Some people do not feel anything, while others may start getting headaches or heartaches right away. Austrian researcher K. Bahler offers a number of reliable signs indicating that a human is living in a geopathic zone."

- difficulty falling asleep

- bad sleep

- anxiousness

- quickened heartbeat

- nervousness and depression

- headaches

- leg cramps

- numbness of legs and arms

- vague pains

- repeated serious illnesses in the same location ("Cancer houses")

- complete and speedy recovery following the change of sleep area

Attention parents!

"A child's organism is the most sensitive to geopathic zones. A child tends to intuitively leave a bad place (geopathic zone). Anxiousness, screams, grinding of teeth, loss of appetite, fear of being alone, desire to get away from bed are the likely signs of geopathic zone effects on a child.

"Despite the fact that geopathic zones have been studied for a long time, until recently, the only true method for their detection had been bio-location (Dowsing) – a method of location of geopathic zones (or underground water) with the help of a rod, a pendulum, or a frame.

"Dowsing clearly determines the main reasons for the formation of geopathic zones – underground waterways, geological anomaly Curry grids, and Hartmann grids. However, the dowsing method is subjective, because it is based on sensitivity of an operator (dowser). Until now, there have been virtually no devices capable of locating geopathic zones. Detection of geopathic zones by measuring geophysical parameters of an area is almost never used, because the measuring equipment is not suited for work in residential or industrial facilities.

"The first portable, electronic device for detection of geopathic zones – Indicator of Geopathic Anomalies (IGA), was developed in 1992. It is a supersensitive receiver of electromagnetic vibrations emitted by a geopathic zone. The IGA device was tested in various conditions for 10 years in many countries. Using the device it is possible to determine location, type, and radiation intensity of a geopathic zone. A sensitive operator (dowser) is not required in order to operate an IGA. Measurements can be conducted in open areas as well as in any buildings, houses, apartments, offices, etc."

4. How to protect from geopathic zone effects.

"The easiest way is to move your sleeping or working place to a safer place, free from Earth's radiation. This practice has been known for some time. About 11,000 people in 14 countries around the world got rid of chronic illnesses simply by moving their sleeping places. Underlining the importance of this simple and effective method of protection from geopathic zones,

Austrian surgeons-oncologists Noth Nagel, Hohenegg, and Dr. Zaurbruch from Germany urge their post operation patients to change their places of sleep and not to return to the same place in order to prevent further destructive influence of the radiation.

"Research and analysis of various existing devices and gadgets for protection against geopathic zones undoubtedly show that complete and reliable protection of a house, as well as the health of its inhabitants from harmful effects of geopathic zones are provided by golden section pyramids.

"Hartmann Grids were discovered in 1950 by a German medical doctor Ernst Hartmann. The lines form a grid around the earth, running north to south and east to west, and extend to a height of 60 - 600 feet. The north-south lines appear approximately every 6 feet 6 inches and the east-west lines appear approximately every 8 feet 2 inches. According to Dr. Hartmann, the worst place that a person can sleep or work is over a Hartmann knot, where two

Hartmann lines cross, as harmful radiation is intensified at this juncture." 10.

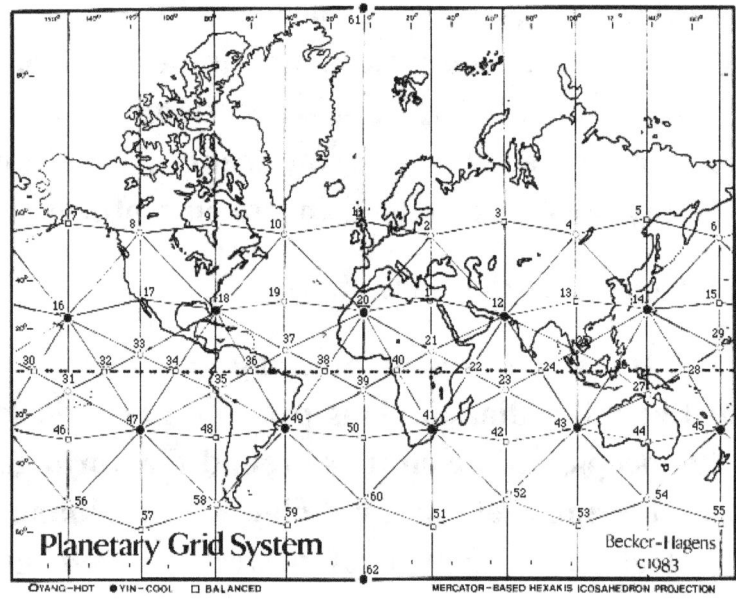

11.

Multi-dimensional Gridding – Deja Allison

"The illustration is an example of how one event can and does affect all dimensions. I have been seeing this for years, now. I have soul traveled the grids since childhood. On these different levels are other manifestations of existence. It may be parallel, or not. I have not drawn ALL the different dimensions and levels available. I could not; it would be impossible to do so. I just gave a sort of quick representation of what layering and dimension looks like as it carries the event.

Multi-Dimensional Grids

Multdimensional Griding

By   Deja Allison

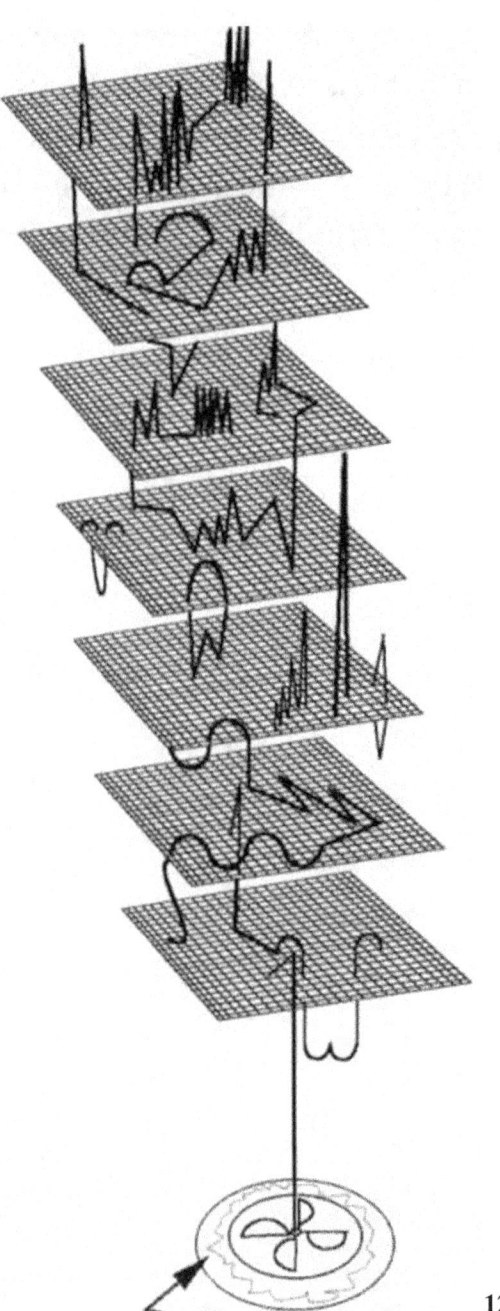

12.

Lotus Petal Pool of Souls

"The bottom part, where you see what looks like a fan is the Pool of Souls. It is the Primal Essence of all creation. It is where ALL is created and manifested from. The squiggle line inside is the best I could do to represent what looks like the petals of an open Lotus flower. I'm not sure I still understand the full potential of that as yet.

"As you can see, no thought or action, no matter if it is past, present, or future escapes the grid. Everything has an effect on everything else. Is this what they mean when they say no person is an island unto themselves? It becomes quite clear why we are responsible for all our thoughts and actions. Look what one thought does! As the effect is moving across the grid it flows down (Only a human directional term for sake of depiction---not an actuality) and enters the pool of continuing flux. This is why we can only create from what we create.

"The universe takes great precaution not to "contaminate" the "Universal Essence of Primordial Flux." This is why there are Gate Keepers as such, for lack of a better way to explain, and there are guardians (again human labeling as a way of explaining) that will not allow entrance. This is also why one can never enter levels or dimension that they are not "frequency ready" to experience. Like begets like energies. It's like a built-in security device. Whatever your frequency is, that is where you will resonate or travel to; that's why most individuals never make it past the Astral Plane. The emotional body must be clear of agenda and negative, human conditioning in order to reach all the many different frequencies available.

"The general rule is: One can travel to the levels/dimensions/universes to which they have the Key or Symbol to---the key being frequency, vibration, resonance, and experience. They may travel there and any that are below that one in frequency, but they can never go higher in frequency than they can achieve at that particular time.

"Could the Universal Primordial Flux become contaminated? Yes, it could. How? By us! If we were to continue to engage in strictly negative thought and action, devoid of any higher spiritual essence, yes, we could in fact contaminate the pool. Eventually, we would create another sub-level of subatomic particles that would become dense matter. How do you think (The word Hell is just a reference word and not meant in a religious context.) what we call Hell was created? It was created by contamination of self-indulgence and negative, ego-based thoughts and actions.

"These create dense energy. I've been reading energy a long time. I can tell you the difference between hatred as a density of energy versus love as a lighter energy manifestation. We are the ones that create our quality of reality---from many different sources. The Earth plane manifestation is a complexity of energies manifested by Space, Time, and Matter."[13.]

The following is an excerpt from renown researcher Gordon Michael Scallion regarding energy grids and their universal influence.

Gordon-Michael Scallion

"Each sub-cell of Earth's energy grid contains the experiences of everything happening, all reality that has ever occurred within it. In other words, each cell retains a complete memory of its entire existence.

"As it relates to areas of population, the memory of what was experienced in a large city, say for example, determines how that city evolves a collective consciousness. The collective karma is affected. The karma of that city, in turn, affects the cell memory karma of that entire region. These memory cells have natural movement due to magnetic I influences. Movement also comes out through human thoughts and actions. No cell is free of karma. Every place has history - moments built upon the moments that came before. Each place is touched by its visitors and inhabitants – not only human but also those physical and ethereal life forms that preceded us.

"Certain cities have developed in a particular way based upon the overall polarity of those cells in which they reside and each city's position on the Earth. So, for example, if we look at those major cities found on or near water, like San Francisco, New York, Los Angeles, Tokyo, or Sydney, we can see that these cities have received great nourishment. These areas have been constantly nourished due to their dynamic positions within the cellular boundaries and in relation to the boundaries of water.

"The balance of elemental forces within each cell affects its makeup, its health, just as the internal thoughts and actions do. Water stimulates that which is intuitive, creative, thus creativity and diversity have flourished in these cities. That was based on the old grid pattern. The energy changes that are now beginning to take place will completely alter the regional dynamics. The new energy will be different.

"The new sub-cells are more numerous and denser in their capacity for memory. So twice as many cells each having individual characteristics will exist. They already exist in the ethereal realm, but the new cells are not receiving very much nourishment from Earth and its elements at present. Most of the energy is still being drawn by the old cells. Right now, the old cells and the new cells are actually competing for energy.

"The old cells are dying, just like skin cells on your body will die. And as they die, they try to draw energy to sustain them, taking it from wherever they can. The environment, the energy of the elements, has been so weakened by our pollution that there is very little left to flow into the cells, old or new. The new cells have remained weak, continuing to draw mostly from the ethereal realm. But with so little energy or food coming from the elements of Nature, the old sub-cells are turning to the only remaining source of vitality compatible with their structure in this reality - man. The cells are taking energy from humans.

"This varies greatly from individual to individual. Just as a city, town, build-ing, or region creates its own karma, so does the individual and collective consciousness of man. How one lives his life determines the kind of interaction between himself and the cell in which he (Or she) lives. Intuitively, they are seeking a more balanced cell. Intuitively, all people know the dynamic be-tween themselves and the Earth. Those who listen and trust are more likely to take action and relocate.

"A balanced person will not draw to himself unbalanced forces. 'Like seeks like.' People in previously energized areas, many of the world's cities for example, could be feeling depleted just by virtue of the fact that they are living in a dying cell, in an environment that does not support rejuvenation. This drain on human energy has been further aggravated in these areas by the dwindling element of air, referring to pollution. Pollution and negative thoughts have reached a point that it is simply no longer possible for this element, air, to rejuvenate the sub-cell region through which it flows. The same condition affects the cleansing properties of the element water - oceans and lakes.

"The new cells operate on a different vibration spectrum not yet touched by our pollution. The new cells are relatively clear of stored memory at this point. It would be comparable to buying a new computer disk. It is ready to hold memory, but until it is put into use, it is only potential. The new cells are not drawing energy from humans. They are much heavier "cosmic-force energy" at this stage. Once they lock in (Which I see in about 3-4 years), as we reach the fulcrum in the processional cycle, they will then seek a rooting, and there will be a period of equalization or balance. This change will happen very rapidly. Suddenly, we will find ourselves in the midst of the new energy, and the old cells will simply be gone.

"The old memories can be retained within the new memory cells and by those individuals and groups attuned to the new cells. But these transferred memories will not have the same intensity or memory charge as they had in the old cells. The storage capacity will increase exponentially! In other words, they could take all the old memory, although perhaps slightly altered and put them all on a little shelf within their total capacity. This is because they have such great volume by comparison. Awareness will function at many levels and dimensions, simultaneously. There will be a much greater capacity for exchange and communication with the Earth force. This is almost like a new software conversion on a computer.

"The invention of computers is nothing more than a manifestation of our collective vision into the future of life itself. Computers were created not to serve man but as an experiment for humans to realize our own true capacity. Computers are a way of trying out our ability to expand our own capacity as humans. This has to do with our changing DNA patterns. There is a transformation occurring in the very blueprint of our makeup. It will be established with the birth of the next root race - the 'Blue Rays.'

"All species, plant, animal, and human are highly adaptable. The basic systems of life, given some minimal parameters, are able to adjust and survive. As Earth's energy cells continue through this change, individuals will be affected in different ways. If you are in a state of balance, meaning that you are working on your life's path, moving forward with little steps that hopefully lead to larger steps, then the conversion and the ability of your body to change need not be that difficult. Everyone will be aware of it to some degree. Electrical sensations in the body will be detectable, even in the most balanced person. Those that are having difficulty in life, however, will need to look at the relationship between themselves and everything else in their lives.

"Stress is one of the things that are going to throw the body's central electrical system even further off-course during these changes. The whole immune system goes right down, because the body needs to adapt to these new grid levels, so it can pull energy from the new pattern. A stressed person will not be able to put the proper internal energy into making the shift.

"So, the new energy will be all around, but the stressed body will not be compatible to receive that energy. Imagine the new cells all emitting energy. The next step for us is to prepare to access the new energy by attuning to it now. This is important for good health in the future. Illnesses that are grid-induced will always be electrical in nature - headaches, fatigue, memory loss, and disorientation, feelings of agitation, vision difficulties, and ringing in the ears. Many of these symptoms may be the result of living in an old cell that is dying, while engaged in a lifestyle that is unhealthy or unbalanced. Most of these conditions can be traced back to a basic imbalance of electrical forces in the body, due primarily to negative thoughts and stress.

"If you are feeling stuck and not moving forward on the truths you seek here on Earth, then you are under stress. This leads to overload and illness. More and more people are showing evidence of being out of balance. More is written about mental illness such as depression these days. Some people are acting violent or erratic for no apparent reason. The energy grid does not cause mental illness. What happens is that the new energy pattern is still grounded in the ethereal realm. It is fluctuating a lot--solar flares, disruptions at the Earth's core, and planetary alignments all influence it, causing the grid to throw up spikes. It is like lightening coming out of the cells.

"When a cell experiences one of these energy spikes, the whole cell actually becomes electrically charged and the sub-cell gets agitated. Their charge level actually increases, which forces a higher charge level in all animals, all plants, and all people within the cell. So now you are taking in Earth force energy at a higher level, at a higher frequency. Most people are taking in pretty low Earth energy and solar energy due to a significant amount of stress. In a state of balance, you can tap into higher psychic abilities. What we think, we shall become. Listen to your intuition (Little voice within)."14.

CHAPTER 5

NON-ATTACHMENT

> He who gives a child a treat,
> Makes joy bells ring in Heaven's street.
> And he who gives a child a home,
> Builds palaces in Kingdom come.
> And he who gives a baby birth,
> Brings Savior Christ again to Earth.
> For life is joy, and mind is fruit,
> And body's precious earth and root.
>
> -   John Mansfield

Release

One of the primary steps in the process of Ascension is that of "being," the releasing of worldly desires. We in the field of metaphysics refer to this as non-attachment. To set oneself apart from desire is to become divine. There is no longer a need to be compulsive over things that one may find absolutely necessary, when, in truth, it was solely a creation in one's own mind.

To explain what is meant to become unattached in your life, allow me to relate a story to you which I call, "The Story of the Perfect Car." There once was a man who bought a beautiful car. In this man's eyes, this car was the most beautiful car that existed. It had everything that you could imagine: great lines, immaculate interior, great horsepower, outstanding stereo equipment, and great resale value.

To make sure that the image of his new car was constantly maintained, the

man would faithfully wash his car twice a week. In fact, he was so obsessed with its cleanliness that, he would not drive it in bad weather. The only problem was, each time he washed his car and drove it around town, he would encounter some situation that soiled his newly washed car. This would happen as a result of some large puddle of water on the street that he had to drive through, or a long stretch of road that was wet and the car in front of him would spray water from the rear tires onto his car, or an automatic sprinkler at a parking lot that would turn on while he was in the store shopping, and the ultimate agitation, driving home late at night on the freeway and thinking that for the first time since he bought the car and washed it, that no soiling would occur. Then, as he was exiting from the freeway and onto the off-ramp, a sprinkler system for the off-ramp ivy plants had earlier turned on and faultily directed its spray across the roadway where he had to drive.

Well, maybe you have never encountered an ordeal such as this to this extreme. I am sure that similar occurrences have crossed your path. This is the problem that happens when we become attached to things in the physical realm, whether it is of an animate or inanimate nature. Life is fleeting, along with everything else in Nature. Nothing lasts forever. Some things stay in your life a while longer than others; however, to believe that your possessions are to remain, regardless of future circumstances, is absurd.

The perfect car did not stay clean but for several hours. The degree of attachment that is placed on an object of desire is in direct proportion to the degree of loss that will occur. When the man with the "Perfect Car Syndrome" gave up his attachment and lost his fear of encountering such situations, the phenomenon ceased to exist.

You can never know what the future has in store. If you think that holding on tightly to all that you have accumulated will bring you happiness, you are living inside an illusory "Perfect Car," inside the matrix of the human mind.

What a combination. Here we are, moment after moment, creating our own perfect, little world within our minds, hoping that all our past experiences will come to our rescue and place us in some utopian lifestyle, free of all pain and misery.

Nature was designed by God, to carry on in a way that abides by a set of axioms. These universal truths are here for everything that exists to obey. When you turn your mind against God's rules, expect to live in a matrix of constant misery. It is only your mind that keeps you a prisoner within its own matrix. Nature is perfect just the way it is. It cannot be any other way except the way it is. God has a Master Plan. We have an amateur plan. Our finite minds cannot compete with the infinitesimalness of Divine Immutability. Everything in Nature appears in direct contrast to God's spiritual realm, but it is only our brain conditioning that creates this illusion. Again, this is the Grand Paradox, One Divine Spirit manifesting outward and observed as the All in everything.

Life is a paradox. What appears to be one thing at times appears as something else at other times. In the case of the "Perfect Car," what was looked upon as bad, water soiling a car was looked upon as good in another situation when the man was thirsty. What we held as truth in the past may appear as a complete falsehood in the future. This is how our mind tricks us. This is what we must understand about life in order to gain happiness. Happiness has always been in our lives; we have just been covering it up with our thoughts. There are two dynamic laws that, when applied correctly, create our reality. These laws were mentioned previously in the book. The first dynamic law is the Law of Association. According to this law, the mind will accept and become that which it is exposed to most frequently. This universal truth applies to all of us. An example of this is the experience of childhood. Depending on the type of beliefs our parents had and expressed around us when we were children, our minds grasp and formulate certain beliefs. As we grow older, our beliefs develop our personalities that we use to move in and about Nature.

Basically, we all have the same needs to sustain life; however, it is our individual imaginations that create the illusions of what we think we need to be happy. The associations our minds hold onto are not necessarily what are needed in our lives; they are merely images of what we think the unknown future should provide for us.

The second dynamic law is the Law of Repetition. According to this law, when you apply your associations numerously over a period of time, those images of association will be impressed upon you through your subconscious mind hologram and expressed in all of your actions. The actions you apply to your life are what cause future circumstances to occur and establish your fate.

Each moment of your life is the only reality there is. The very last moment is gone forever and becomes a part of the past, and the next moment has not arrived yet, so, it is an illusion created by your mind of what you think should happen. The universe exists right now. This very moment is as real as it gets. To live for each moment, one at a time, is to create a mindset that will bring you all the happiness you can ever hope to imagine.

Associate your thoughts with a present time mentality. Repeat this mindset, one second at a time, until your life is filled with experiences where you focus on what is before you. Put your future goals where they belong, in the imagined future. It is only when you place your full energy on the present that the future eventually arrives to become the present and bring you abundance and joy.

This entire non-attachment issue is very tricky. Even though it is the necessary thing to do to achieve happiness, it is practical to say that it is virtually impossible to maintain 100% of the time. However, when you practice holding onto the present and avoid drifting into the realms of past and future, the degree to which your life will experience equilibrium and miraculous events will astound you. You are truly the master of your universe, the universe of your "True Self." All the answers are within you. Free your mind. Free the irrational impulses that can cause your emotions to erupt.

Stand aside from your thoughts. Detach yourself from them completely. They are not really who you are. Know that when thoughts enter your mind, they are merely energy impulses. These are nothing more than images imposing themselves upon your consciousness. You can never know what thought or image will appear next; they are activated by any number of impulses. Remember the computer "pop-ups" analogy in the previous chapter. When harmful thoughts arise on the surface of your mind, remain neutral and unbiased. Practice will manifest a diminishing of their influence.

Nature is constantly placing images and objects before you that can influence you. This is why it is important to stay alert in present time, to control and give direction to the thoughts that enter your internal environment. Once you recognize thoughts for what they truly are, energy brain stimulations that have been stored in your subconscious mind hologram and ready to manifest themselves into your conscious mind, those unwanted messages automatically lose 50% of their power! From then on, the degree of their influence diminishes.

Find a place where you can go every day and be in a state of silence. This is a period of reflection that is so important for your evolution. It is at these moments that you can enter the inner silence of your being and connect with your spiritual self. It is then, that you can unveil your true calling from God, and then express your soul to the world.

Ascension

"The three billion base pairs in the human genome are organized into 24 distinct, physical, separate, microscopic units called chromosomes. The total genetic information within these structures is a sequence of over 6 billion base pairs or 12 billion "letters." That makes up only 3% of what is known. Another 97% remains unknown to conventional science. Spiritual attunement is attuning oneself to a higher pitch.

"Ascension means to rise up. An ascension experience is a period of time in which your material body raises to a higher state. It is nothing new, as others throughout time have experienced this. Perhaps what is new about 'this time' is that it is happening to people 'en- masse,' contributing to a greater spiritual awakening of consciousness. Again, one group of a highly energetic body has a tendency to raise all those in the vicinity. Our physical body is actually becoming a different frequency (Component of) of light as we go from a more material body to a more refined light-body. This is a physics principle in action.

"The Ascension places you in the midst of universal secrets hidden from veiled eyes, and so the more important your wisdom and power become. As we become more interdimensional and multidimensional, the frequencies alter, and the energies accelerate; your body goes through a drastic, rapid change that the nervous system must handle. Some signs and symptoms reported are: rejuvenated organs, hot/cold flashes, massive dizziness, change of diet, excessive tiredness, skin, hair, and eyes changing, increase of aches and pains, increased psychic and kundalini activity, etc. Many have read and/or experienced such (And continue to do so).

"Ascension appears to be a biological reality; there is an evolution of species occurring on this planet, as well as an evolution of consciousness. Many sensitive and attuned individuals feel and see the change happening within them and subsequently receive the same Spirit messages from the cosmic grid of consciousness, or as Carl Jung termed it, 'the collective unconscious.'

"This metamorphosis includes very specific biological and bio-chemical changes in your DNA, cells, blood, endocrine system, and brain, thus affecting the energy of the body throughout the dimensional planes. All of these changes are taking place on the level of sub-atomic particles."[15].

Dr. Berrenda Fox provides evidence of DNA and cellular changes and has given "ample" proof through blood tests that some people have actually developed new strands of DNA. Mutations and molecular changes are happening. Hopefully more of the scientific community comes forward with their findings to substantiate the information that is there.

Harmonic Convergences

"We have also all heard in spiritual circles of the harmonic convergences and light grid activations, time for the consciousness of mankind to make a quantum shift to the next level of experience. These shifts and activations are merely the harmonizing of those atomic pieces that make up the whole. The cosmic ocean, as it is called throughout the world, generates cosmic waves of energy. The ocean consisting of our thoughts, feelings, emotions, and intents: love, hate, anger, etc. ride upon and make up these waves, which in one way can be compared to the 'quantum foam.'

"The "Harmonic Convergence" is not just a New-Age trekkie term for the fun of merely saying it, because it sounds cool. It is a real 'quantum' event. We cannot discount that this is really happening on a level we cannot physically/visibly see. According to physics principles, this affects every single organism in the universe and on Earth.

"Dropping polarity - no longer thinking in terms of good or bad but seeing the wholeness and completion and perfection of life. It is a very definite mind, emotional, and body response. The body response is the one where your body simply does not acknowledge good or bad but sees that there is a higher purpose behind it all.

"We must remember our true multidimensional nature so "we" can merge into the multi-dimensional consciousness that is our innate heritage. All the dimensions are ever-present and functioning in and/or through the physical body and throughout the universe. Due to our dense vibration rate (Right now), we are not aware of them until we somehow remember and recall our multidimensional awareness."

DNA

"UC Davis experts on basic DNA research. Since 1953, DNA research has had an impact on everything from biology, agriculture, and medicine to criminal law, justice, art, and politics. DNA structure is more than a double helix. The famous double helix is actually supported by many different proteins that allow it to be read, copied, and repaired. Working out what those proteins are and how they work together is the aim of the Center for Genetics and Development led by Stephen Kowalczykowski, professor of microbiology and molecular cell biology. Kowalczykowski's own laboratory studies the family of proteins that allow DNA strands to cross over and recombine, creating genetic variation."[16].

Science of DNA

"We can apply the concept of frequency and vibration to the body and its cells as physics dictates. We can also study the body and its building blocks and infer the same information applies to DNA.

"As in the infamous Fibonacci numbers, DNA and the body can also be converted to mathematical, physio-musical signals. Professor Susumu Ohno of the Beckman Research Institute of the City of Hope proposed years ago, that the repetition process governs both the musical composition and the DNA sequence construction. The body creates music through brain waves, the heartbeat, blood circulation, endocrine cycles, and right up to the microwave level of organ vibration. DNA is the orchestrator of this music.

"Molecular biologists have defined 20 proteins in the body which they have been able to look at music and read the exact sequence of bases in the DNA. One part of the DNA structure when "played" from the sequences sounds like a waltz melody, another sounds similar to an Irish jig, with a definite beginning, a middle, and an end. Oddly enough, the base sequence T-T-T-C-C-C-C-C-C-, when played, are the famous opening notes of Beethoven's Fifth Symphony!

"It (The music) is produced from the primary and secondary structure of protein sequences....pitch is determined by amino acid identity, and instrumentation is chosen according to a protein folding pattern with different instruments representing regions of alpha-helix, beta-strands and turns. The fundamental point that it is not is that everything that exists has fields of vibrations, but that everything that exists is fields of vibrations."[17].

CHAPTER 6

SYNCHRONICITY

Victories that are cheap are cheap.
Those only are worth having which come as a result of hard fighting.

-     Henry Ward Beecher

Synchronicity-Gears/Wheels of Time 18.

Synchronicities are people, places, or events that your soul attracts into your life to help you evolve or to place emphasis on something going on in your life. The more consciously aware you become of how your soul creates, the higher your frequency goes and the faster your soul manifests. Each day your life will become filled with meaningful coincidences and synchronicities that you have attracted or created in the grid of your experiences in the physical. There are no accidents, just synchronicity wheels, the gears of time, the wheels of time, the wheel of karma-wheels within wheels, the alchemy of creation, the Philosopher's Stone, Sacred Geometry=SG=Star Gate, evolution of consciousness.

Do be careful. Not all synchronicities are positive. Sometimes these lead to learning lessons when you are deceived into thinking that it is the road to take at that moment in time. This is not always the case, so do be careful. If you are a dysfunctional person, a drama person, you will attract and manifest dysfunctional people and events.

Synchronicities can also go nowhere, as they just occur in someone's life to make a point. You must look at the bigger picture of the synchronicity. Think

outside the box, not at the actual event. Look at the underlying facts when the synchronicity occurs, to be sure you know why you attracted that person/situation into your life.

Criticism

Since the theory of synchronicity is not testable according to the classical, scientific method, it is not widely regarded as scientific at all, but rather as pseudo-scientific or an example of magical thinking. However, it is doubtful that Carl Jung would have considered the theory to be scientifically testable.

Probability theory can attempt to explain events such as the plum pudding incident (Explained later) in our normal world without any interference by any universal, alignment forces. However, the correct variables required for actually computing the probability cannot be found. This is not to say that synchronicity is not a good model for describing a certain kind of human experience, but according to the scientific method, it is a reason for the refusal of the idea that synchronicity should be considered a "hard fact," i.e., an actually existing principle of our universe.

Supporters of the theory claim that since the scientific method is applicable only to those phenomena that are reproducible, independent of the observer and quantifiable, the argument that synchronicity is not scientifically 'provable' should be considered a red herring; as by definition, synchronistic events are not independent of the observer, since the observer's unique history is precisely what gives the synchronistic event meaning for the observer.

A synchronistic event appears like just another meaningless, 'random' event to anyone else without the unique, prior history which correlates the event. This reasoning claims that the principle of synchronicity raises the question of the subjectivity of significance and meaning in the sequence of natural events. The feeling of making connection where there is none is called apophenia.

Alternative Explanations

Aspects of the subjective experience of schizophrenia have much in common with the subjective experience of synchronicity, in the sense that ordinary events are seen as having a direct, personal relevance to the schizophrenic but are seen as 'normal' by non-schizophrenics. Many psychoses are similar to schizophrenia but can last for a very short time, such as in rare instances from nicotine withdrawal, causing the same effect even with a non-schizophrenic. Those who have experienced a near-death experience or kundalini awakening report an increase in synchronistic events happening to them. A religious analogy of this experience might be attributed to the fulfillment of prayer or miracles; however, Jung did not describe it in these terms.

Correlation can also be described as an 'actual connecting principle' and so has been proposed as an analogy to the phenomenon of synchronicity. Though correlation does not necessarily imply causation, yet correlation may have in fact been a physical property shared by events without there being a classical cause-effect relationship, as shown in quantum physics, where widely separated events can be correlated without being linked by a direct physical cause-effect.

Synchronicity has been proposed as a corollary phenomenon of the many-worlds or parallel universes theory of quantum physics, in that the subject is somehow 'navigating' to those particular alternative worlds that are correlated in their past history, among the myriad possible other worlds that are not correlated to their past history. Although this idea has made it into the popular press, it is considered pseudo-science by most scientists as the parallel universe theory states that all possible futures exist simultaneously, therefore, the subject indeed lives out all possible futures in parallel.

A well-known example of synchronicity involves plum pudding. It is the true story of the French writer, Emile Deschamps, who in 1805 is treated to some plum pudding by the stranger Monsieur de Fontgibu. Ten years later, he encounters plum pudding on the menu of a Paris restaurant and wants to order some, but the waiter tells him the last dish has already been served to another customer, who Examples of Synchronicity turns out to be Monsieur de Fontgibu. Many years later in 1832, Emile Deschamps is at a diner and is once again offered plum pudding. He recalls the earlier incident and tells his friends that only Monsieur de Fontgibu is missing to make the setting complete, and in the same instant the now senile Monsieur de Fontgibu enters the room by mistake.

You meet someone who interests you and touches your soul. Through synchronicity, that person seems to come into your life over and over again.

You begin to feel a destiny with that person. You begin to think with your heart instead of your head. You connect with that person. In some cases the karma between the two people is positive, but in many cases you have attracted that person into your life for a learning lesson, whether you are aware of it or not.

• You can consider an event synchronistic when an inner experience such as a dream, vision, or other form of déjà vu prepares you for the physical event. They are in your life when financial difficulties seem to have no end, yet there is always enough money for basic expenses, rent, food, utilities. Finances seem to appear where and when they are needed.

• You have just received your last check from unemployment, when suddenly, a job comes along.

• You walk into a bookstore, not knowing what to buy, and the book you need falls from a shelf and practically hits you over the head.

• You have been feeling ill with no apparent cure. You are out for the day and meet someone who knows a doctor or healer with the answers.

• There is a sudden relocation which seems to be for one reason, and you find much more than you bargained for.

• You finally end a bad relationship and immediately another partner comes into your life.

• You feel depressed and can't find focus in your life, and the next person you talk to says something that brings you the guidance you need.

• Everyone's favorite…..You drive to a place where parking is 'next to impossible,' and someone pulls out of a parking spot, or it is just waiting for you.

Strange things happen to everyone. Science cannot interpret them in a logical or reasonable manner, and therefore they are regarded as mere coincidences. For example, an old friend calls you up a minute after a distant recollection flashes through your head; somebody pays you back when you desperately need money. There are coincidences that are more complex and even more astounding. They make you think about the inexplicable and divine intervention. How do they happen contrary to laws of Nature and science?

A ship sunk near the shores of Wales on December 5, 1664. Another ship went down on the same location on December 5, 1785. A third ship sunk to the bottom of the sea on the same location on December 5, 1866. Each of the shipwrecks had only one survivor. In each of the three cases, the survivor's name was Hugh Williams.

There is another story of amazing coincidence relating to the fate of U.S. president John Kennedy and U.S. president Abraham Lincoln. The simplest coincidence has to do with the surnames of the two presidents. Both surnames are made up of seven letters. Abraham Lincoln was assassinated in a theater; his assassin tried to hide in a warehouse. Lee Harvey Oswald fired his rifle from a position in a warehouse; he tried to hide in a theater. President Lincoln had a secretary named Kennedy. President Kennedy had a secretary named Lincoln.

On a hot, July night in 1930, a policeman named, Allan Folby, from Texas got in a car crash and damaged the femoral artery in his leg. He would have bled to death had not a passerby intervened and saved his life. The passerby's name was Alfred Smith. He applied a tourniquet and Folby stopped bleeding. The policeman recovered and went back to his beat. Five years later, he got a call to proceed to a car crash site. He saw a man lying on the ground. The man was bleeding; his femoral artery was apparently damaged. The man's name was Alfred Smith; he was the very man who saved Folby's life in the past under the same circumstances.

Why do such circumstances take place? Many people believe such events are in fact made on purpose to make a person take a more careful look at his or her life and interpret its events in a more serious manner. Just like somebody sends us a message that reads: trust your intuition. Maybe the cosmos (God) gives an impulse or a clue: believe in coincidence.

The latter assumption may seem unscientific; however, let us not forget that some events we call "coincidental" other people have been calling the "phenomenon of synchronism" for a long time. Two outstanding scientists came up with the term regardless of each other. Nobel Prize winners in physics, Wolfgang Paulli and Carl Gustaf Jung, a famous psychiatrist and psychoanalyst arrived at the conclusion that simultaneous occurrence of causally unrelated events has meaning beyond mere coincidence. Those events may occur without any reason or respect to existing laws of Nature. The laws of Nature are not absolute said the scientists while explaining their positions.

Both Jung and Paulli maintained that relations between the events can be of fundamental meaning to man even though the belief is in contradiction to the fixed concepts of the universe and its principle laws.

Carl Jung

Synchronicities are Meaningful Coincidences

Synchronicity is a word coined by the Swiss psychiatrist, Carl Gustav Jung, to describe the temporally coincident occurrences of acausal events. It was a principle that he felt compassed his concept of the collective unconscious, in that it was descriptive of a governing dynamic that underlay the whole of human experience and history, social, emotional, psychological, and spiritual.

Jung believed that many experiences perceived as coincidence were due not merely to chance but instead potentially reflected the manifestation of coincident events or circumstances consequent to this governing dynamic. Jung spoke of synchronicity as being an "acausal connecting principle" (i.e. a pattern of connection that is not explained by causality).

Psychiatrist, Carl Jung believed the traditional notions of causality were incapable of explaining some of the more improbable forms of coincidence.

Where it is plain, Jung felt that no causal connection can be demonstrated between two events but where a meaningful relationship nevertheless exists between them, a wholly different type of principle is likely to be operating. Jung called this principle, "synchronicity."

In, "The Structure and Dynamics of the Psyche," Jung describes how, during his research into the phenomenon of the collective unconscious, he began to observe coincidences that were connected in such a meaningful way that their occurrences seemed to defy the calculations of probability. He provided numerous examples culled from his own psychiatric case studies, many now legendary.

"A young woman I was treating had, at a critical moment, a dream in which she was given a golden scarab. While she was telling me her dream, I sat with my back to the closed window. Suddenly, I heard a noise behind me, like a gentle tapping. I turned around and saw a flying insect knocking against the window pane from outside. I opened the window and caught the creature in the air as it flew in. It was the nearest analogy to the golden scarab that one finds in our latitudes, a scarabaeid beetle, the common rose-chafer (Cetoaia errata), which contrary to its usual habits had evidently felt an urge to get into a dark room at this particular moment. I must admit that nothing like it ever happened to me before or since and that the dream of the patient has remained unique in my experience."[19].

Who then, might we say, was responsible for the synchronous arrival of the beetle, Jung or the patient? While on the surface reasonable, such a question presupposes a chain of causality Jung claimed was absent from such exper-ience. As psychoanalyst Nandor Fodor has observed, the scarab by Jung's view had no determinable cause but instead complimented the "impossibility" of the analysis. The disturbance also (as synchronicities often do) prefigured a profound transformation. For, as Fodor observes, Jung's patient had, until the appearance of the beetle, shown excessive rationality, remaining psycho-logically inaccessible. Once presented with the scarab, however, her demeanor improved, and their sessions together grew more profitable.

Because Jung believed the phenomenon of synchronicity was primarily connected with psychic conditions, he felt that such couplings of inner (Subjective) and outer (Objective) reality evolved through the influence of archetypes, patterns inherent in the human psyche and shared by all human-kind. These patterns, or "primordial images," as Jung sometimes refers to them, comprise mans' collective unconscious, representing the dynamic source of all human confrontation with death, conflict, love, sex, rebirth, and mystical experience. When an archetype is activated by an emotionally charged event (Such as a tragedy), says Jung, other related events tend to draw near. In this way the archetypes become a doorway that provides us access to the experience of meaningful (And often insightful) coincidence.

Implicit in Jung's concept of synchronicity is the belief in the "oneness" of the universe. As Jung expressed it, "Such phenomenon betrays a "peculiar inter-dependence of objective elements among themselves as well as with the subjective (Psychic) states of the observer or observers."[20]. Jung claimed to have found evidence of this interdependence not only in his psychiatric studies but in his research of esoteric practices as well.

Of the I Ching, a Chinese method of divination which Jung regarded as the clearest expression of the synchronicity principle, he wrote, "The Chinese mind, as I see it in the I Ching, seems to be exclusively preoccupied with the chance aspect of events. What we call coincidence seems to be the chief concern of this peculiar mind, and what we worship as causality passes almost unnoticed. While the Western mind carefully sifts, weighs, selects, classifies, and isolates, the Chinese picture of the moment encompasses everything down to the minutest, nonsensical detail, because all of the ingredients make up the observed moment."[21].

Similarly, Jung discovered the synchronicity within the I Ching also extended to astrology. In a letter to Freud dated June 12, 1911, he wrote, "My evenings are taken up largely with astrology. I make horoscope calculations in order to find a clue to the core of psychological truth. Some remarkable things have turned up which will certainly appear incredible to you. I dare say that we shall one day discover in astrology a good deal of knowledge that has been intuitively projected into the heavens." [22].

Freud was alarmed by Jung's letter. Jung's interest in synchronicity and the paranormal rankled the strict, materialistic Freud; he condemned Jung for wallowing in what he called the "black tide of mud of occultism." Just two years earlier, during a visit to Freud in Vienna, Jung had attempted to defend his beliefs and sparked a heated debate. Freud's skepticism remained calcified as ever, causing him to dismiss Jung's paranormal leanings, "in terms of so shallow a positivism," recalls Jung, "that I had difficulty in checking the sharp retort on the tip of my tongue."[23]. A shocking synchronistic event followed.

Jung Writes in His Memoirs

"While Freud was going on his way, I had a curious sensation. It was as if my diaphragm were made of iron and were becoming red-hot, a glowing vault. And at that moment there was such a loud report in the bookcase which stood right next to us that we both started up in alarm, fearing the thing was going to topple over on us. I said to Freud that that was an example of a catalytic, exteriorization phenomenon. "Oh, come," he exclaimed. "That is sheer bosh." "It is not," I replied. "You are mistaken, Herr Professor.

And to prove my point, I now predict that in a moment, there will be another such loud report! Sure enough, no sooner had I said the words that the same detonation went off in the bookcase. To this day I do not know what gave me this certainty, but I knew beyond all doubt that the report would come again. Freud only stared aghast at me. I do not know what was in his mind or what his look meant. In any case, this incident aroused his distrust of me, and I had the feeling that I had done something against him. I never afterward discussed the incident with him."[24].

In formulating his synchronicity principle, Jung was influenced to a profound degree by the "new" physics of the twentieth century, which had begun to explore the possible role of consciousness in the physical world. "Physics," wrote Jung in 1946, "has demonstrated that in the realm of atomic magnitudes, objective reality presupposes an observer, and that only on this condition is a satisfactory scheme of explanation possible. "This means," he added, "that a subjective element attaches to the physicist's world picture, and secondly that a connection necessarily exists between the psyche to be explained and the objective space-time continuum."[25].

These discoveries not only helped loosen physics from the iron grip of its materialistic world view but confirmed what Jung recognized intuitively, that matter and consciousness, far from operating independently of each other are in fact interconnected in an essential way, functioning as complementary aspects of a unified reality.

The belief suggested by quantum theory and by reports of synchronous events that matter and consciousness interpenetrate is, of course, far from new.

• Synchronicity reveals the meaningful connections between the subjective and objective world.

• Synchronistic events provide an immediate religious experience as a direct encounter with the compensatory patterning of events in Nature as a whole, both inwardly and outwardly.

As you now have observed, much research has been conducted in this area of synchronicity. It is of further interest to know that related to this phenomena of synchronism is a study which penetrates the very fabric of universal essence and could provide even more evidence as to how events that occur seemingly coincidentally are in fact interconnected. Chapter 7 will project you into the realm of interrelated realities and unveil strategic evidence to an all oneness throughout the universe.

CHAPTER 7

NON-LOCALITY

> When wealth is lost, nothing is lost.
> When health is lost, something is lost.
> When character is lost, all is lost.
>
> - Anonymous

Non-Locality, Defined

To explain this phenomenon, let us look at the discoveries of scientists who are continuously researching these new realms, so that you may draw your own conclusions and even expand upon the existing material.

Non-Locality, action at a distance, is the concept that an object can be moved, changed, or otherwise affected without being physically touched (As in mechanical contact) by another object. That is, it is the nonlocal interaction of objects that are separated in space. Pioneering physicist Albert Einstein described the phenomenon as "spooky action at a distance."[26].

"Albert Einstein's general theory of relativity opened the doors of science along with the mystical realities. Einstein theorized that space and time intertwine, and that matter is inseparable from an ever-present quantum energy field, and this is the sole reality underlying all appearances. This theory challenged the basic assumptions about the universe and what it contained.

Physicists found that the most basic atomic particles in the cosmos comprise the very fabric of the material universe. An electron can be shown to be both a wave and a particle, depending on the observer's perspective.

Physicist, David Bohm, in his plasma experiments at the Berkeley Radiation Laboratory, found that individual electrons act as part of an interconnected whole. In plasma, the gaseous electrons more or less assume the nature of a self-regulating organism, as if they were inherently intelligent.

This scientific discovery of Non-Locality, the wave/particle duality, meant that everything is joined or connected together. Space and time is composed of the same essence as matter. Bohm found this to be a conscious, atomic sea and extending out from this sub-atomic reality all of material creation may also be said to be conscious. Since all matter and events interact with each other, time, (Past, present, future) along with space and distance, all is relative to the observer and operate as one under the law of Non-Locality.

A principle related to Non-Locality is called Bell's Theorem. This is a quantum physics law that says that once connected, objects affect one another forever, no matter where they are. Following the principle of Bell's Theorem, an invisible stream of energy will always connect any two objects that have been connected in any way in the past, meaning that everything connects to everything else and that physical reality is BOTH waves and particles. This model birthed the "Holographic Universe" idea, the powerful conscious energy that the whole can invariably be found in the tiniest particles: an atom of a blade of grass to the most distant galaxies."[27]. "The building blocks of atoms are merely, "parcels of compressed energy, packed and patterned according to certain mathematical formulae."[28].

"Princeton Engineering Anomolies Research Program at Princeton University refers to quantum concepts such as the principles of complementarity (The action in one system can affect the actions of another system at a quantum, energetic level, free of the limits of time and space) and wave mechanical resonance (Matter and energy exchanging manifestations as vibrating waves and particles)."[29].

Matter and Energy are two poles of the same unity. Shamans and Mystics call this Oneness or Interconnectedness.

"I believe a leaf of grass is no less than the journey work of the stars; also. I am large; I contain multitudes."30. Walt Whitman

Michael Talbot, in, "The Holographic Universe," describes all of material creation as a "ripple...a pattern of excitation in the midst of an unimaginably vast ocean, and despite its apparent materiality and enormous size, the universe does not exist in and of itself but is the stepchild of something far vaster and more ineffable."31.

Bohm believes that our almost universal tendency to fragment the world and ignore the dynamic interconnectedness of all things is responsible for many of our problems, because we believe we can extract the valuable parts of Earth without affecting the whole...treat parts of our body and not be concerned with the whole...deal with crime, poverty, and drug addiction without addressing society as a whole.

In the book, "Stalking the Wild Pendulum," Itzhak Bentov outlines sound and vibration wave behaviors, atomic molecular structure, realities, and time, along with the quality and quantity of consciousness and the various levels of our realities. "We know that reality is made up of two components, one an immutable line-background and the other dynamic, which are a vibrating aspect of the same thing. Then we know that mind and matter are made of the same, basic stuff." Restated, "We compare solid matter to ice, while comparing mind or consciousness to steam or vapor, all being the same, basic stuff in different form."32.

The Kaluza-Klein Theory was the first theory of higher dimensions. It simply states the gravity in which light could be explained as 5th dimensional vibrations. This evolved into, "Super Gravity Theory," which then led physicists to the, "Superstring Theory;" it postulates that all matter consists of vibrating strings. This theory raised the standards and levels of mathematical science, because it seems to explain all the fundamental laws of Nature.

"The symmetries of the sub-atomic realm are but remnants of the symmetries of higher, dimensional space."[33].

Michio Kaku says in, "Hyperspace," "Physicists....are now seriously studying multiple connected worlds as a practical model of our universe."[34].

George Bernhard Riemann was the first to lay the mathematical foundation of geometries in higher, dimensional space and birth the idea of a "Simultaneous 4th Dimension." He claimed, "Universes are completely self-consistent and obey their own logic."[35]. He tried to discover, "the unity of all physical laws," which appears simple when expressed in higher dimensional space. In this process, he redefined Euclidean geometry.

The importance and significance of mathematics and the profound role we have come to understand within Nature can also be found in the Fibonacci numbers. This logarithmic series of numbers demonstrates the occurrence of many analogous, spiral forms in Nature. This underlies fractal mathematics and allows imaging of our natural world. Fibonacci shapes are found in the shape of galaxies, the nautilus shell, and the double helix of the DNA molecule, to name a few.

Unification Theory

The "Unification Theory" is built on the fundamental knowledge of physics, the oneness and interrelationship and Nature: ecologically, individually, and the whole world. The knowledge of this system and its dynamic capacity to survive, grow, and transform in relation to the other dynamic systems is most profound. The theory encompasses all living and non-living systems that exist, which parallels the particle and wave duality and its coexistence within the whole universal system of operation. Countless mathematicians, physicists, philosophers, and religious thinkers have attempted throughout the ages to find the one thing that can tie all things together. See page 9.

Consciousness

With the entire breakthrough in the dynamics of our natural world, the topic of physics and consciousness is becoming more well-renowned by physicists. In the spring of 2003, the Quantum Mind Conference on "Consciousness, Quantum Physics, and the Brain" was held in Arizona, USA. Their website states, "Recent experimental evidence suggests quantum Non-Locality occurring in conscious and subconscious brain function, and functional quantum processes in molecular biology is becoming more and more apparent. Molecular biology is becoming more and more apparent. Moreover, macroscopic quantum processes are being proposed as intrinsic features in cosmology, evolution, and social interactions."[36].

Perhaps in knowing we are all part of this Non-Locality or Oneness, we can make strides to improve our society, our nation, and our world.

"A scientist can have an almost approaching religious experience, realizing that we are children of the stars, and that our minds are capable of understanding the universal laws that they obey."[37].

Telekinesis

The year was 1980. I was studying for my masters and doctorate in metaphysics. At the time, I was living in a ground-floor apartment in Mission Beach, a beach community in San Diego, California. Six months previously, I had just ended a romance with a young woman whom I was very much in love with. I missed her terribly and attempted to bring her back into my life. It just so happened that I was studying the science of telepathy and preparing a paper on the subject for one of my university courses. I had read an article in a newspaper about a psychic who uses telekinesis (The ability to control objects or thoughts from a distance) for certain Hollywood actors who wanted to influence the mind of a selected producer in order for them to secure a lead role in a motion picture. Well, I was a little skeptical when I read of his powers, but I figured that I could attempt two tasks at the same time, complete my study of telepathy and win back my lady.

I telephoned the psychic in Las Vegas and told him of my situation. He stated that he would need a personal article of the person I was attempting to bring back. I told him that it had been six months since I last saw her, and I doubted that I had anything of her personal belonging, but I would search my stored containers for any remnants. Luckily, I discovered a small beret she wore in her hair. I phoned him two days later; he said the beret would do. When I questioned his technique, he stated that he would lie in a hot bath and relax while envisioning thoughts being transferred to this particular person. He mentioned holding the beret while performing this technique. The person in question left a vibration in the inanimate object, which he was able to sense and therefore communicate across the grid that permeates all physical world objects. I asked him how long it would take to achieve results. He told me to give him a week.

Approximately six days later, I was preparing a sandwich in my kitchen. It was early evening; the sun had just ended another day. Since my apartment was ground level and on the oceanfront walk, I drew the drapes of the large, bay window every evening in order that I might enjoy some privacy. As I was finishing preparing the sandwich, I thought I heard something against the bay window. The stereo was on; I figured it was the music. Then, I heard a tapping sound on the window. I walked over to the window and opened the drapes. Much to my surprise, my lady friend was standing there, smiling. At that very moment, I became a true believer. These powers lay dormant within the confines of our own minds. We are born with innate talents that can be surfaced, nurtured, and applied for our advantage.

These phenomenons are all a part of our ongoing investigation into the unseen realms of universal vibration. In Chapter 6, "Synchronicity," the "Bi-Mass/Velocity Theory" in conjunction with the "Equation of Human Events" plays the primary role in determining synchronistic events. The collective consciousness of all human beings, past and present, is available throughout our energy grid. Even though similar events are sometimes many years apart, the momentums of human beings (Again based on decisions derived from intellect, emotions, desires) can create a series of similar events formulated by the enormous quantity of possible scenarios in the "causal vault." I agree that the laws of probability are a factor; however, I believe we can journey outside these laws and consider the phenomena that have been discussed in this book's chapters.

CHAPTER 8

HOLOGRAPHIC UNIVERSE AND INTERCONNECTEDNESS

> Live as if you were to die tomorrow.
> Learn as if you were to live forever.
>
> - Mahatma Gandhi

In his bestselling book, "Think and Grow Rich," Napoleon Hill stated that, someday someone would scientifically reveal how we receive our consciousness through the Ether. My research has revealed this evidence and is explained in my book, "The Spiritual Big Bang." Amazon.com

Akashic Records 38.

This is a theosophical term referring to a universal filing system that records every occurring thought, word, and action. The records are impressed on a subtle substance called akasha (Or Soniferous Ether). In Hindu mysticism, this akasha is thought to be the primary principle of Nature from which the other, four, natural principles: fire, air, earth, and water were created. These five principles also represent the five senses of human beings.

Some indicate the akashic records are similar to a Cosmic or collective consciousness. The records have been referred to by different names, including Cosmic Mind, Universal Mind, and the collective unconscious (Collective subconscious). Others think the akashic records make clairvoyance and psychic possible.

It is believed by some, that the events recorded upon the akasha can be ascertained or read in certain states of consciousness. Such states of consciousness can be induced by certain stages of sleep, weakness, illness, drugs, and meditation, so not only mystics but ordinary people can and do perceive the akashic records. Some mystics claim to be able to reanimate their contents like they were turning on a celestial television set. Yogis also believe that these records can be perceived in certain psychic states.

Certain persons in subconscious states do read the akashic records. An explanation for these phenomena is that the akashic records are the macrocosm of the individual subconscious mind. Both function similarly; they possess thoughts which are never forgotten. The collective subconscious gathers all thoughts from each subconscious mind hologram, which can be read by other subconscious minds.

An example of one who many claimed successfully read the akashic records is the late, American mystic, Edgar Cayce. Cayce did his readings in a sleep state or trance. Cayce's method was described by Dr. Wesley H. Ketchum, who for several years used Cayce as an adjunct for his medical practice, "Cayce's subconscious is in direct communication with all other subconscious minds and is capable of interpreting, through his objective mind and imparting impressions received, to other objective minds, gathering in this way all knowledge possessed by endless millions of other subconscious minds."38. Apparently, Cayce was interpreting the collective subconscious mind long before the psychiatrist C.J. Jung postulated his concept of the collective unconscious.

On April 21, 2017, I watched the latest edition of Ancient Aliens entitled, "Ancient Aliens: Declassified," History Channel. I have been watching the Ancient Aliens series since it began and they finally reached the part, where they are talking about cosmic connection and how humans receive knowledge. They featured some of the greatest minds: Tesla, Einstein, Newton, Jobs, Ramanujan, who claimed to have experienced some sort of out of world influence that produced the creative thoughts that manifested into great, scientific discoveries. Ancient alien supporters claim that the influences came from aliens manipulating human consciousness and providing inspirations to certain individuals that would guide us along in our consciousness evolution. The thought entered my head, "Well, it doesn't appear that they are doing such a great job; look at what's going on in the world."

What many of my science colleagues still haven't come to the realization of is that consciousness and inspiration do not originate from alien beings, if they do exist, and even if they do exist, we would discover that other species of life received their consciousness in the same manner that we do, through universal frequency.

In India, they believe in the Akashic Records that claim to hold a record of all action that has taken place since the creation, all the knowledge of one universal mind; we are a collective of that mind. These are encouraging claims, but just how does this system of One Universal Mind scientifically function. No one has ever charted and connected all the dots from origin to insertion of our consciousness, until now.

This is the inspiration I received through a One Universal Mind, but I had to establish validity for my claim. This took many years of scientific research and vision, to enter realms of thought, reason, logic, and imagination that were unique only to me. It is a lonely space when you are on the perimeter of new discovery, and no one else understands the concept. Henry Ward Beecher once said, "The philosophy of one century is the common sense of the next." Anyone who has ever patented an invention has felt this in his/her life.

The Akashic Records claim is a valid claim; however, it needs revision. No one ever described and illustrated any details about this universal phenomenon. What does it look like; where is its origin; how, scientifically, does it connect to our consciousness? These are all questions that needed to be solved. The inspirations I received set forth a new paradigm regarding God's spiritual location within all creation and how it is God's universal frequencies that are the Akashic Records; it is God's frequencies that are responsible for our evolution of consciousness; it is God's frequencies that carry the <u>coding</u> for all preexisting knowledge available to humanity. God created the knowledge that establishes the rules for science and math; the universe created by God functions in accordance with His own laws. We receive this knowledge through God's universal frequency. Wouldn't you say that we need a conclusion episode on The History Channel featuring this evidence?

Early Childhood Experience

Long before I received degrees in metaphysics, I began contemplating unusual events that occurred in the physical world. An incident that started it all took place when I was six years old. My best friend and I were discussing dreams one day, and we discovered that we both dreamed the same dream the night before. I thought he was joking with me, but when I asked him to tell me some of the details, he had the answers.

The dream entailed the two of us walking through a jungle, when suddenly we were attacked by natives. We ran as fast as we could to the bridge that stretched across the river. The natives were throwing spears at us, but we knew that if we could get to the bridge and stand on it, even penetrating spears could not harm us. We made it to our destination.

My parents told me of an experience involving the three of us, that I found overwhelming. I was born during World War II. In 1944, my mother was pregnant with me and having a very difficult time with labor. Her condition worsened to the point that the Last Rights of the Catholic Church may be administered to her. My father was in the Navy and stationed at Norfolk,

Virginia Naval Base. His naval convoy was scheduled to leave for the Mediterranean Sea the next day. Word of my mother's condition reached him, and he was given a leave of absence, upon which he returned to Pennsylvania.

Upon my father's arrival at my mother's side, her condition improved, and I was born. Simultaneously, my father's convoy left port without him; however, unfortunately, the ship my father would have been on was torpedoed and sunk while crossing the Atlantic Ocean. All crew members were lost.

You will recall in Chapter 3, your introduction to grids. It is my knowing that a large grid composition permeates the entire universe. This grid of sub-atomic mesh wiring that I know originates from the source of God's dimension within all physical world phenomena is, in essence, the basis for karmic cause. Albert Einstein had the same knowing about a grid; however, he did not reveal its origin. Karma is the path of all action (Momentum) in the physical universe. Every single motion that occurs sends a vibration into its immediate environment. These waves range in intensity from subtle to tremendous. When a wave of action is committed, that action is imprinted into the cosmic fabric of the grid, forever.

I am sure you have heard of the saying, "What goes around comes around." This is the law of karmic cause. After the imprint of an action has been established, the wheel of karma carries it until a situation in your life, as a result of the "Bi-Mass/Velocity Factor" and the "Equation of Human Events" (Chapter 1), merits its return action into your life. From this interpretation, it is simple to understand that no action goes undetected. You might say that this could be regarded as God's Universal Law Enforcement Agency (GULEA), all-knowing, all-seeing, all-engaging.

It is my belief that those who are gifted in the realm of psychic, clairvoyant, and telekinesis have the ability to see into this fabric and view the actual event that has been logged into the grid. All the events that have ever vibrated are instilled with permanency into God's eternal memory bank of physical world events.

Dynamics of Thought

"We are what we think. All that we are arises with our thoughts. With our thoughts, we make our world."39.

Thought is a great, dynamic force with tremendous power. A simple atom radiates an electromagnetic field or EMF. A molecule radiates yet a stronger EMF. A large number of molecules form a cell in which the human brain contains at least 200 million such cells. These atomic waves seem to include the emotions and are thought to be superimposed within the EMF frequencies. These vibrations are sent into the physical world, extending outward into the "non-physical" world. The historian Arthur Koestler refers to this as, "The capacity of the human psyche to act as a cosmic resonator."40.

Thoughts paired with intent equals a very powerful mixture of, "mental stuff." Thoughts and emotions happen as a result of your interactions with the world based on your perception of past experiences, of how you perceive this world of events that form your thought processes that become reactions and behaviors, unconsciously or consciously. Your thoughts create your feelings, emotions, behavior, and what you attract and magnetize into your life.

 In minding your thoughts, there is also a Japanese principle known as the, "Kototama Principle," meaning "word soul." When one develops the right connection to language and words, this deep connection becomes a 'mirror' against which one reflects concerns in order to have a better sense of one's nature, place, and relationship to a totality. This "hidden" philosophy is just now re-emerging as our understanding of how thought and feelings affect our realities.

Emotion is the fuel of creation...... Everything you dwell upon in your mind and heart, everything that you believe in is drawn into physical manifestation. This is happening more and more quickly, as the masters have advised humankind for thousands of years.

"Thank goodness for feelings that create our emotions. Feeling and emotion are what will ignite your desire to know the love that is within you."[41].

The Brain

"The human brain, being formed of an inestimable number of spherical resonators, is termed in medical science as nerve cells forming the gray matter of the brain. These minute spheres make up the thought force that permeates all space in endless waves, eternally active. This force we term atomolic; the cells are composed of atomoles, whose vibratory motions under the action of universal thought force result in the phenomena of thought, cognition, intellection, etc. Understanding this, no one should continue to feel surprised at the caring emotions and impulses of a human being in an undeveloped state, as only by developed WILL can the motions of this force be directed."[42].

Additionally, quantum wave function of the brain is not completely random and seems to have a "phase difference," which implies the parallels of mind and body to the duality of the wave and particle. The wave aspect of Nature yields the mental; the particle aspect of Nature yields the material.

For centuries, Buddhists have believed that pursuing such practices seems to make people calmer, happier, and more loving. At the same time, they are less and less prone to destructive emotions. Mindfulness meditation strengthens the neurological circuits that calm a part of the brain that acts as a trigger for fear and anger. This raises the possibility that we have a way to create a kind of buffer between the brain's violent impulses and our actions.

Laughter

We have all heard that laughter is the best medicine. The emotions that come from humor and laughter help people cope better with the stress of daily life. Studies have shown that people who laugh often live longer and are much healthier compared to those that find it difficult to laugh due to life's circumstances and personality types. This common reflex affects the body by activating the cardiovascular system, helps lower blood pressure, and also exercises the muscles of the body. Research has shown that parts of the limbic system are involved in laughter as well. The limbic system is a primitive part of the brain that is involved in emotions and helps us with basic functions necessary for survival.

Dr. Lee S. Berk at Loma Linda University School of Medicine and Public Health in California and endocrinologist Stanley Tan studied various groups of adults and found that laughter also stimulates our endocrine system and the pituitary gland, which in turn stimulates release of endorphins and encephalon, natural painkillers that are chemical cousins to opiates such as morphine and heroin. They found that both arms of the immune system got a boost out of laughter. Other studies have shown laughter improved functioning in those with breast cancer, marital stress, and also those in the grieving process.

Silvia Cardoso, a behavioral biologist reports, "Repeated, short, strong contractions of the chest muscles, diaphragm, and abdomen increase blood flow into our internal organs and forced respiration (The Ha! Ha!), assures this blood is well oxygenated.

Muscle tension decreases, and indeed we may temporarily lose control of our limbs, as in the expression, "weak with laughter." It may also release brain endorphins, reducing sensitivity to pain and boosting endurance and pleasurable sensations. Some studies suggest that laughter affects the immune system by reducing the production of hormones associated with stress and that when you laugh, the immune system produces more T-cells."43.

Recent surveys indicate that laughter can enhance the quality of our conversations, productivity, and social interactions. It simply makes people feel closer to each other. In his book, "Laughter: A Scientific Investigation," Dr. Provine says that, "Laughter is the oil in the social machine, helping human interactions run more smoothly."44. Laughter sounds also have emotional responses in listeners. Recent studies support the notion that one important function of laugh acoustics is to influence the emotional responses of listeners, the researchers from the Acoustic Society of America meeting conclude.

I had stated in the Frequency and Sound portion that the universe consists solely of waves of motion. "There exists nothing other than vibration," relates Walter Russell in, "A New Concept of the Universe." "In vibratory physics, the principles that make sound into harmonious music are the same principles that govern all associating vibrations throughout the universe - and that includes everything there is. Vibrations are dynamic things, not unlike "living" things, since they are in a mutual state of "harmony."45.

David Bohm summarizes the M=E=I formula (matter = energy = information) when he writes: "There is a limitless amount of information enfolded into the structure of the universe, and we are a manifestation of that energy. Every body event, whether the workings of enzymes, neuro-peptides, hormones, blood, or skin, is an info-energetic event."46.

"Science has discovered startling, new possibilities regarding how we think, feel, love, heal, and find meaning in our life...research suggests that the heart thinks, cells remember, and that both of these processes are related to an as yet mysterious, extremely powerful but very subtle energy. Science may be taking the first steps to understanding more about the shamans/healers/-leaders – the energy of the human spirit and the coded information that is the human soul."[47].

Many cardiologists have combined their knowledge of biology and new physics - the study of subtle energy and the invisible atomic world and modern cardiology. Energy cardiology is based on the, "Dynamic Systems Memory Theory," the idea that all systems are constantly exchanging mutually, influential energy which contains information that alters the systems taking part in the exchange. From the, "Heart's Code," website, "Our billions of heart cells are able to form a quantum energy generator, by laws of the quantum world such as non-locality, the info-energetic connection that constitutes all systems in the universe."[48].

Another field, Kinesiology, has made strides in determining feelings and emotions on the well-being of the body. David R. Hawkins, MD, PhD conducted a 29 year study that demonstrated the human body becomes stronger or weaker, depending on a person's mental state using kinesiology or muscle testing. Dr. Hawkins developed a scale of 1-1000 that maps human consciousness. 200 (or 20,000 cycles per second) weakens the body and from 200 to 1000 makes the body stronger. At various points above this are intellect, psychology, science, genius, heart, and enlightenment, which move us from duality to non-duality. This Map of Consciousness is based on a logarithmic scale, which means there is an enormous increase in power with even a small increase in the number on the scale.

The lowest vibration rate is described as "Force." This exhibits a weak kinesiology response with a vibration under 200, which includes shame, then guilt, apathy, grief, fear, anxiety, craving, anger, and hate. All these feelings weaken you. Levels at 200 and above are described as "Power," which elicit a strong kinesiology response. At 250 you have neutrality and trust - this strengthens you.

Going up on the scale there is: willingness, optimism, acceptance and forgiveness, reason and understanding, love and reverence, joy and serenity.

Courage calibrates at 200

Willingness calibrates at 310

Reason calibrates at 400

Unconditional Love calibrates at 500

Joy and the ability to communicate healing energies to another calibrates at 540

Gratitude calibrates at 560

Praise at 570

Peace and bliss is rated at 600

Enlightenment and Ineffability at 700-1000.

Dr. Steven Hawkins conjectures the great scientists such as Newton and Freud calibrated at 499, and Carl Jung calibrated at 540, the highest level of intellect we have known to this point.

At the highest end of the kinesiology scale of human experience, 600 to 1000 are qualities that transcend duality. This is the level of "enlightenment" and "pure bliss" of consciousness.

The highest calibration in this study was 700 by Mother Theresa, who exuded peace and joy by her mere presence. Dr. Hawkins reports that Jesus, Krishna, and Buddha, the greatest prophets recorded in human history all calibrate at a scale at 1000, along with their messages of unity, peace, and love calibrating near 1000.

Cycles Conversion of Hawkins's scale illustrate:

Hawkins's Scale	Cycles P/Second= Hz	KHZ or kc/s
100	10,000	
200	20,000	
300	30,000	
500	50,000	
1000	10,000	

The frequencies from light and sound to physiological functions (As demonstrated in the Frequency section), along with bio-scientific studies show definite resonances in all things and ultimately show the dual nature of our physical being and how we are energetically linked to everything else.

"Energy waves build the human body by radiation in intrauterine life as a pattern of energy current and continue to maintain it by this energy flow as wireless currents."[49].

Prayer

To "feel" combines our emotions merged with our thoughts; energy follows attention. When we pray, we realize and acknowledge there are many possibilities, and by the act of praying-focusing-feeling, we direct that energy into an opportunity that we "choose" to experience - of all the possibilities available.

A Tibetan abbot said on prayer, "Feeling - the object of each prayer is to achieve 'feeling.' When we pray, we feel on behalf of all beings everywhere. We are all connected. We are all expressions of one life. No matter where we are, our prayers are heard by all. We are the entire, same One."[50]. "Unless you yourself enter the image and think from it, it is incapable of birth."[51].

Gregg Braden and others have spoken about the power of prayer and the physics principles at work. "Through the language of time waves, quantum outcomes, and choice points, we determine our future probabilities. Our individual choices merge into our collective response to the present, whose rings span far into the depths of time/space."[52].

The matrix grid of consciousness, Jung's collective unconscious, is the nervous system for the Earth. Our thoughts affect this system and everything contained therein. The universe can be crafted with a simple working of your will. Mages and mystics and shamanic healers have taught this truth throughout the ages. Thoughts and feelings manifest themselves into reality. This makes us co-creators with the universe. Quantum physics even suggests that by re-directing our focus and our attentions, we can bring a new course of events into action.

"Everything we think, feel, and do has an effect on our ancestors and all future generations and reverberates throughout the universe." [53].

A tiny emotion becomes a thought of one raindrop; thousands of drops turn into a flood. There is true magic-alchemy in the power of will; we shape the world by our desires, thoughts, intents, and actions. So, in relation to mankind's "ascension," it is the product of many working, thinking, acting, and living as a higher unit of collective thought, energy processes.

Kaballah teachings have a saying, "Think well of everyone." In Kaballah, what we do in this world always has effects in the spiritual dimension. There is also the term, 'cycle of reciprocity,' what is going out from your mind is coming back to you. Dhyani Ywahoo, a Cherokee spiritual teacher, says it beautifully. "Be aware of the power of mind; remember that we are all in the process of unfolding."

We can choose; we can weave; we hold the form; we dance it, and the moment comes when it is recalled in each of us...By the power of its sound, it is a reality...The Beauty Path, the Great Peace is the meeting of ourselves, the perceptions of our minds and the cessation of those waves and thought forms that create discord. Let us sow the seed of peace in all our actions, thoughts, and words."[54].

"The Spiritual Path assists mortal man in gaining and understanding his/her place within the Divine Order and allows mankind to experience the greater purpose of evolution from a more universal perspective."[55].

"When mankind perceives reality through an open, evolved, heart chakra, mortal man experiences reality by 'feeling' the energy states of that reality. We can safely say that real truths cannot be said, talked about, or even written; they must be lived, experienced, and felt."[56]. "Remember to be cautious with your thoughts and your words; they are the tools of creation. Remember cause and effect. Remember there is nothing more important than love."[57].

In conclusion, when the claim that our universe is a hologram, it is important to realize that, when there is a spiritual essence within all physical world phenomena, there can be no holographic universe. A hologram manifests as the result of a photographic technique that records the light scattered from a 3D object with mass, splits the beam, projects it onto a photographic plate, and then presents it in a way that appears three-dimensional (Chapter 1).

God's Pure Light, (No mass) inner dimension cannot create a holographic universe spatially beyond Itself. This is in direct contradiction to what constitutes a hologram. Holograms are created from a three-dimensional object, as the subconscious mind hologram is a hologram of the brain. Scientists who believe that our universe is a hologram would have to believe that God resides and created our universe from a fourth dimension, making our physical universe a hologram of the fourth dimension. If that be the case, why are they searching for the God Particle deep <u>within</u> our physical universe?

There is a science that defines prayer. When prayers are performed, the prayer asked for is visualized at the pre-frontal cortex (3rd Eye) of the brain. The brain sends it to the subconscious mind hologram, where the necessary data is collected, translated, stored, and then relayed back to the brain for application as decision-making and problem-solving solutions. The subconscious mind hologram supplies the brain with data that will situate you along a path that brings you into contact with those forces (X factors that you have no control over) that are necessary to manifest your prayer into reality. For thousands of years, when people prayed and the prayer came true, they claimed it as a miracle, and it was. They just did not realized God's science at that time and how the collective consciousness functioned.

CHAPTER 9

UNIVERSAL FREQUENCIES

I'm gonna fly away and catch the sun.
 Hold it in my hands where there used to be none.
 And when it fills me with its golden rays,
 I'll share it with you till my dying days.

- Damon Sprock

The Wave Vibration 58.

Walter Russell states, "To know the mechanics of the wave is to know the entire secret of Nature."

The psychodynamics of the mind as an electromagnetic structure establishes the nature and reality of consciousness as an interdimensional energy process. It is an electrical process of cause and effect.

Consciousness determines our vibration (Frequency and amplitude). When we realize that time and space is really ONE, I think we then go beyond a simple understanding. This is what Einstein was talking about concerning frequency (Time) and amplitude (Energy)...and time traveling - and this would mean being spiritually energized to walk in the 4th dimension, freely. Physics states the 4th dimension exists (Subsequently physicists say 10 dimensions exist that they have mathematically figured so far); we can thus apply it to other spiritual areas, as in various mysticism practices. I disagree and explain on page 127.

Scientists say that when our energy-amplitude is raised higher, it is synonymous with our frequency (Time) getting higher, and when our frequency raises, the distance between the waves are closer, therefore shortening time considerably. So, if time is the distance between two points, what your amplitude-energy is determines how efficient you travel that time line. They say, for instance, you are standing by me.......you raise your energy, which raises your frequency (If you are efficient at it, efficient in science is the least amount of work for the energy expanded.) You dart across the room and come back to standing by me. I do not see you - except for the transition period of your energy, you coming and going. You might look like a fuzz of light at first, (Dart across the room and come back) then a fuzz of light, again. This is all that would be noticeable. Or perhaps, if one is in movement while this energy transition is taking place, we may observe a 'blurring' and see many, different phases as is seen in someone taking consecutive pictures of a track and field long jumper. The realities are 'blurred' together, and as physics states, can be seen in the multiple dimensions at once.

This parallels the concept of the "Superstring Theory," which assumes there are an infinite and alternate series of parallel universes that exist. The viewpoint of many theoretical physicists: light is a vibration of the unseen 4th dimension. What is light? Light, electricity, and magnetism are manifestations of the same thing called electromagnetic radiation that encompasses the electromagnetic spectrum in which only a small fraction of light is visible to us. When we do the "dart across the room" experiment, only the small, visible light will be seen as a fraction of the whole, energetic transition.

At this point is where I have to intervene. When you observe how the universe was created by a spiritual entity from an internal, infinite, irreducible, spiritual dimension, it will be clear that the assumption that multiple universes exist and that the 4th dimension allows one to walk freely is invalid. The natural progression in creation places the higher dimensions of consciousness into the inner dimensions – God's infinite, internal, irreducible, spiritual dimension. This is in direct contradiction with the present belief that dimensions beyond our third dimension are sources of higher consciousness.

Dimensions beyond our third dimension would be of greater density, not lesser density. Each of the three dimensions that we are familiar with progress in density. The furthest inner dimension that science has been able to detect with scientific devices is the sub-atomic particles, the so called "God Particle" being the latest scientific claim. They are the lightest density, because they are closest to Origin, God's most inner, pure light dimension. The second dimension, the atom dimension, is of greater density than the sub-atomic particle dimension, and the third dimension, all the forms we see around us, is of greater density than the atom dimension. The natural progression after that would be a fourth dimension with a greater density than the third dimension, and so on and so forth.

Forms in the fourth dimension would be made up of particles that are more closely compact, causing the forms to weigh greater than the same forms of the third dimension. Forms in the fourth dimension would be so heavy that momentum would be slowed down considerably. This is my conception of HELL, where souls are sent for atonement, to experience excruciating mental and emotional pain as a result of the high density.

The inner dimension of spiritual consciousness is the source of our human consciousness, a quantum connection that makes available how we receive information through universal frequencies and how the frequencies form our subconscious mind hologram, the inner, vibrating dimension of the brain that contains all preexisting potential.

When we resonate, it can also be viewed as "musical notes" that represents a mode of vibration, a distinct resonance or particle. Michio Kaku, a physicist and author of, "Hyperspace," states. "Matter is nothing but the harmonies created by this vibrating string. The laws of physics can be compared to the laws of harmony allowed on the string. The universe itself, composed of countless vibrating strings, would then be comparable to a symphony."59.

When sound harmonics form direct harmonic relations, the two vibrating sounds and their chords of vibration are said to be sympathetic to each other. In other words, resonance - like attracts like. This combined frequency dictates that what happens to one is vibratory and happens to the other, simultaneously, in varying degrees of harmony or dissonance. This is the Law of Cycles as presented by John W. Keely's, "Law of Sympathetic Vibratory Physics.

"Music is formless and is therefore the perfect expression of the formless."60. Musical notes or sounds can be converted numerically as in the example in the previous section regarding the Fibonacci numbers found in Nature. According to mathematics principles, a series of vibrations will multiply many times its own pitch, thus increasing harmonically all those that are near. This is the Law of Harmonic Pitch.

This multiplication of frequency and amplitudes is thought of in relation to the Ascension process. All things are energy, thus our "energy bodies" are becoming more attuned, reaching a higher vibration rate to the dynamic, energetic force of the universe. Extending and expanding vibration rates happen in Nature, and we are merely a part of Nature. So, we should, too, in theory, be capable of" Ascension as dense, physical, material beings.

Sound

"He made the whole world an instrument of sound and an instrument for carrying messages, resounding praise to the Creator of all."61. Frank Waters – "The Book of the Hopi." There is something called the Sacred Sound Current which correlates to the spiritual Hindu "OM" and the Sufi "HU" and the Sant Mat "Shad." This Music of the Spheres, Logos, or Word as called in mystical teachings goes beyond our limited, spoken language. Such ethereal music is referred to also as the Sound of God or the Breath of God since the sound permeates all things. And as we know, all things are composed of energy.

This all-pervading sound in the current of the Ocean of Consciousness is believed to be the connecting link between God and man. The divine currents are always playing a symphony of ethereal music; however, it does not become audible until we have progressed spiritually enough to hear with the inner ear. Additionally, many mystics, ancient and modern, have reported seeing various colors with the inner eye(s) while in meditation.

As physics deems, all is in a vibration state and thus can be converted into mathematical renditions, musical notes, and color frequencies. Shad is believed to be the pinnacle point of all the powers of Nature in which, according to scriptures, is the method in which God creates.

"I dwell in undefiled Light, and a Thought revealed itself perceptibly through the great Sound. And it is a Word by virtue of the Sound; it was sent to illumine those who dwell in darkness. I am a Light that illuminates the All. I am the Light..."[62].

Light and Sound is also viewed as synonymous with each other in the ancient scriptures. This has been mentioned in various contexts in this writing as well. They are Interconnected and part of the Whole within the wave/particle duality.

We can parallel the holographic universe concept with Philo of Alexandria circa 40 CE, "The whole creation, this entire world perceived by our senses, is a copy of the Divine Image. But the shadow of God is His Word which He used like an instrument when He was making the world. And this shadow, as it was, model, is the archetype of other things."[63]. Also, similarity to Carl Jung's archetypes within the collective unconscious.

"The musician is very close to mysticism, far closer than the philosopher, because music is meaningful without any words; it is meaningful simply, because it rings some bells in your heart, creates a synchronicity between you and itself when your heart starts resonating in the same way, when you start pulsating in the same way."[64].

Frequency

There are known quantitative frequencies that have been measured that illustrate the body and energy connections with comparative measurements to the world around us.

^ Means to the power of 10, ^-8, 10 the minus 8 power

Gamma Rays 10^{18} to 10^{26}hz

X rays 10^{16} to 10^{20}hz

Ultraviolet Light (UV) 10^{15} to 10^{17}hz

Visible Light $10^{4.7}$ to 10^{15}hz

Infrared Rays 10^{11} to $10^{4.7}$hz

Radar Waves $10^{9.5}$ to $10^{11.5}$hz

Microwaves 10^{8} to $10^{9.5}$hz

Television Waves 10^{6} to 10^{8}hz

Radio Waves $10^{1}$to $10^{6.5}$hz

Elf Waves 10^{1}hz

Earths Ionosphere Cavity Resonance - Schumann Field ranges from 1 Hz to 30 Hz - Radiation Spectrum

Audio20   20,000 Hz

Ultrasonic   20,000 Hz - 100-kHz - 10 MHz

Radio Frequency 150KHz-1.5 MHz

High Radio Frequency RF   1.5MHz-40 MHz

Very High Radio Frequency VHF   40MHz-100MH

Ultrahigh   Radio

Frequency UHF   over 100MHz

Brain:

Alpha   8-13 cycles/sec - relaxed alert

Beta   14-30-intense mental concentration, agitation

Theta4-7 Hypnogogic

Delta 5-3.5 Sleeping

Human Tolerances to sinusoidal vibration

Head Pain   13-30Hz

Impaired Speech 13-20Hz

Jaw Pain   6-8Hz

Chest Pain 5-7Hz

Abdominal Pain   4.5 - 10 Hz

Human Body Cell1,520,000- 9,460,000 Hz

Upper Limit of Human Hearing   15,000 Hz

Molds, viruses, bacteria, worms, mites range from 77 KHz to 900 KHz

Warts   400-430 Kz

Tapeworms   420-450 KHz

Mites   640 KHz - 850 KHz

Ant   1000 to 1200 KHz

Goldfish   900 to 1500 KHz

Chameleon 1000 to 6000 KHz

Cat   1500 to 8000 KHz

"Biological systems are influenced by the terrestrial, electrical environment. This environment includes electric fields, magnetic fields, field modulations, aerion (positive and negative) concentrations, and biological waves. Nature ordains harmonious variations of these factors. The human organism exhibits revealing electrical characteristics.

"Electromagnetic brain waves (0.1 to 30 Hz) occur at frequencies paralleling those of terrestrial spherics and the Schumann resonance. Decision making abilities are subordinate to alpha, beta, gamma (.01-5hz), theta (4 - 7hz), and delta (0.1-4hz) brain rhythms with their related states of consciousness."65. Symposiums on Biological Effects and Measurement of Bio-frequency/ Microwaves, Department of Health, Education and Welfare, FDA – Critical Aspects of Human Versus Terrestrial Electromagnetic Symbiosis, E. Stanton Maxey, M.D. , 1977.

Duality

As you have learned thus far, all life co-existing together at the same time is matter and space, movement and stillness, sound and silence, light and darkness. As we have seen earlier, the wave/particle duality (Non-Locality) means that everything is joined or connected together. This duality also means that objects that appear to have seemingly opposite properties are of the same substance but in a different form or density of a vibration level.

It is believed that Earth in what is deemed as a "free-will zone," is the ongoing battle between the positive and negative forces, the dark vs. light. Scientifically speaking, there is no difference, just differences in frequencies and amplitudes. Simplistically speaking, we are all one.

"Light and shade, long and short, black and white can only be experienced in relation to each other; light is not independent of shade, nor black of white. There are no opposites, only relationships."[66].

Another example of this polarity principle is found in sound harmonics. Fields may be enharmonic on one level but harmonic on different levels. Enharmonic would be like plucking a guitar string and allowing it to vibrate freely. The sound and the frequencies of the corresponding vibration are no longer exactly in the same frequency ratio. Harmonics is the vibration multiples of the same frequency.

The Taoists also viewed the dynamic interplay between the polar opposites as yin and yang. This pair of opposites shares a polar relationship, where the two poles are dynamically linked to the other and found throughout Nature.

This idea of implicit unity of all opposites can be extremely difficult to accept. However paradoxical it may seem to some, Eastern philosophies have always considered this as essential for attaining enlightenment, to go 'beyond earthly opposites,' and in China, the polar relationship of all opposites lies at the very basis of Taoist thought. In the Tao of Physics, thus Chuang Tzu says, "The 'this' is also 'that.' The 'that' is also 'this.' . . . That the 'that' and the 'this' cease to be opposites is the very essence of Tao. Only this essence, an axis as it were, is the center of the circle responding to the endless changes."[67].

This ancient wisdom of duality is the essence of the Tao. It also appears to be the essence of physics. "Every explicit duality is an implicit unity."[68].

Heraclitus, the Greek philosopher, also realized that all opposites have a unity and relativity. "The way up and down is one and the same and, God is day-night, winter-summer, war-peace, and satiety-hunger."[69].

Science and religion also represent a complimentary pair of polar opposites residing within a highly complex paradox. If we view these opposites as the ends of the light spectrum, we have a diverse range of human culture and ideals. These cultural dimensions in this range quantify the ideal of "unity of diversity," biologically, ecologically, and sociologically. In, "The Wisdom of Thich Nhat Hanh," he eloquently says. "Realizing the interdependent nature of dust, flowers, and humans, we see that unity cannot exist without diversity. Unity and diversity interpenetrate each other fully."[70]. The full magnitude of the paradox disappears in the light of diverse duality. "Life starts with the knowledge of diversity, but the awareness of unity is the pinnacle of life."[71].

The Kybalion

The Kybalion encompasses the study of, "The Hermetic Philosophy of Ancient Egypt and Greece." This is considered as the life work of Hermes Trismegistus, who was considered to be the 'scribe of the gods.' This Greco-Egyptian philosopher and mathematician reportedly had written tens of thousands of volumes of books revealing to humankind the healing arts, magic, writing, astrology, science, chemistry, and philosophy. This is a time of unity between the scientific and religious studies. Hermes also built several pyramids and reportedly even taught the great mathematician, Pythagoras. Oddly enough, Hermes is considered the "Great Central Sun Occultism." Perhaps Hermes was the living embodiment of duality, as he inscribed some of the earliest accounts of the secret of the ultimate reconciliation between metaphysical science and religion. His wisdom follows in excerpts from the Kybalion. The parallels to "modern "thinking are astounding. "Everything is Dual; everything has poles; everything has its pair of opposites; like and unlike are the same; opposites are identical in nature but different in degree; extremes meet; all truths are but half-truths; all paradoxes may be reconciled."72.

"The same Principle manifests in the case of "Light and Darkness," which is the same thing, the difference consisting of varying degrees between the two poles of the phenomenon."73.

The 'Art of Polarization' becomes a phase of 'Mental Alchemy' known and practiced by the ancient and modern Hermetic Masters. An understanding of the Principle will enable one to change his own Polarity, as well as that of others, if he will devote the time and study necessary to master the art.

The teachers claim that illustrations of this Principle may be had on every hand and from an examination into the real nature of anything. They begin by showing that Spirit and Matter are but the two poles of the same thing, the intermediate planes being merely degrees of vibration. They show that the ALL and the MANY are the same, the difference being merely a matter of degree of Mental Manifestation. By looking carefully at the nature of opposites, we might realize that each pole actually incorporates its opposite in one form or another as an essential component of the whole.

"To understand the whole, it is necessary to understand the parts. To understand the parts, it is necessary to understand the whole. Such is the circle of understanding."[74].

CHAPTER 10

THE NEW HUMANS

Wake up all the teachers, time to teach a new way.
Maybe then they'll listen to what you have to say.
They're the ones who's comin' up, and the world is in their hands.
When you teach the children, teach them the very best you can.

- Harold Melvin and the Blue Notes

"The Children/Adults Who Fell Through the Cracks in Our Society"

By: Laura Lee Mistycah 75.

Where It All Started

It has come to my attention that more than 85% of the "True Knights" on this planet today (which includes women) are "First Wave Indigo Children/Adults." They have filtered into this planet for over 70 years, coming en masse (about 62%) between the years of 1969 and 1987, and about 30% were born in the 50s.

Here is a brief profile of what a First Wave Indigo is all about and also some of the challenges they have had to overcome. Hopefully this information will be of benefit and give some relief to you gallant Indigos out there who have been searching for answers as to why your world is so different, so challenging, and in many cases, "So Hellish!"

The First Wave

Lately, we have been hearing so much about "Indigo Children".... newborn babies, toddlers, and children who seem to have what I term "Hyper-Intelligence," mentally, psychically, and spiritually. These children are making their statements loud and clear…."I Am Here"…."Pay Attention to Me!"….and people around them are responding to their voice.

But let's go back to where it all started taking shape, back to the 60s, back to the time of political, spiritual, and sexual revolution, back to a time when the Earth took a major, energetic shift…...a time when our society's "rebels of the world" were starting to have children of their own, a time when the music changed.............when Pink Floyd, The Moody Blues, The Beatles, The Doors, Iron Butterfly, and Led Zeppelin created a new paradigm of lyrics and tone expression….where an explosion of self-induced drugs took people to spaces, places, and dimensions that are not of this world. (And sadly, some of them never returned.) Nothing would ever be the same again.

This is where things started taking form for the Indigo Ray.

These are Beings from another octave, same tones but a higher octave…a higher vibration…a higher awareness, and an entirely different perception! The ways of the lower octaves, though functional for the lower octave scale, just plain do not work for the higher octave beings, and the more you try to make them conform to the lower octave scale, the more resistance and frustration you get from both sides because,…it just will not "fit" nor "comply."

These Beings are very brilliant and psychic, but our current, scholastic, evaluation systems do not include tests that show the "true colors" of these extraordinary Beings, and they are being labeled as: "Hard to teach," "slow," "dyslexic," "uncooperative," "quiet," "A.D.D.," and even "retarded." The sad outcome of our inability to understand and cultivate the ultra-intelligence of these higher octave Beings is that they have been ridiculed, rejected, abandoned, and outcast, with many of them becoming loners and reclusive from time to time, because there is nowhere for them to go but "inside."

The truth of it all is that these phenomenal Indigos have been sent here by their volition, to change the entire structure of the planet's thought system known as the Consciousness Grid. Our outdated rigidity embedded in this plane, with its heavily-protected ignorance, is being abruptly transformed by those who embody the "Indigo Ray!"

Below are just some of the qualities and challenges that First Wave Indigos experience. Most Indigos will relate to at least 90% of the list.

- First Wave Indigo Profiles were born en masse between 1969 and 1987 (With stragglers before and after)

- Highly intelligent in their "Own Way."

- Are literally "wired" differently than other people.

- Many have strong or unusual Psychic and Telekinetic abilities.

- Can relate well to children and/or the elderly.

- Involve themselves in human/animal rights efforts (Which include the plant & elemental kingdoms.)

- Indigos have a strong bond/connection to the trees and Nature. Many times get along better with animals and Nature than with other people.

- Creative, inventive, and very intuitive.

- Have an innate sense of "Oneness" and connectedness to all of Creation.

- Have an innate sense of "Oneness" and connectedness to all of Creation.

- Get confused and disturbed when others don't share their reality of "at-one-ment."

- May go through periods of apathy and cynicism as coping mechanisms. Intense longing for those of their own kind"....Soul Mates.......but don't know where to look.

- Have what I endearingly term H.D.D. "Hug Deficit Disorder" and need immense amounts of physical touching, hugs, and love to "cuddle."

- Because of being misunderstood and then betrayed, may develop strong trust issues and therefore keep many of their thoughts, feelings, and opinions to themselves.

- About 30% have difficulties expressing themselves, especially in writing. NOTE: If you read some of the poorly written correspondence from some of these First Wave Indigos, you would assume they were uneducated and nearly illiterate, but the truth is these same people can also be speed readers and can absorb information in seconds that would take others minutes to understand and retain.

- Very disciplined when properly motivated.

- Get bored and/or frustrated in school.

- Male Indigos (And many Female Indigos) for the most part don't "do authority" very well, because most of the time they are smarter than those in authority.

- Many find themselves in "Alternative Schools."

- Female Indigos seem to be able to tolerate and cope better with the school systems than their male counterparts.

- Have a strong sense of truth, ethics, justice, and freedom. (That is why "authority figures" many time s irritate and frustrate them.) When these attributes are in jeopardy, Indigos will give their "all" for their cause and many times feel they would rather die than give-in to tyranny and deception.

- Many are labeled "Dyslexic" and find themselves in "Special" classes at school that usually never work for them.

- Indigos have a strong desire to know "why"…and if they don't see "the point" in something, (Or if is it isn't explained properly) will feel it is not not worth their time/energy and will either react with resistance or just simply "blow off" the people/things that they deem unworthy of their efforts.

- Innately have their own way of doing calculations, usually in their head; many are accused of cheating, because they cannot show their work.

- Indigos have an evolved awareness of how things work. Therefore, many of the rigid rules and methods of learning Math, English, and Physics (Not Metaphysics or Quantum Physics) make no sense to them.

- All First Wave Indigos have what might be termed "A Gift of Healing….. Whether it is making people feel better with their humor and wit, hands-On healing, animal and plant healing, healing with music and tone, or healing with new, "unproven" methods….some of which are natural and need no external training for. Most Indigos have "Telepathic Healing" abilities, and long distances make no difference to the efficiency of their work.

- Because of their expanded perception, unusual creativity, wanting to try new things, and running way ahead of what is being taught in class, many were diagnosed as having Attention Deficit Disorder and put on Ritalin as children.

- Most Indigos (Especially males) have a high, innate aptitude for computers/electronics and/or machinery/auto mechanics. It is common for them to "just know" how to operate and trouble-shoot with very little help from a book, manual, or instructor.

- First Wave Indigos are extremely creative and express this innate skill in many different and often times outrageous forms. These skills manifest in: drawing, painting, sculpting, decorating, photography, writing (In sometimes very extreme and unique ways). Making blueprints and prototypes, composing and playing music (even if they have never had lessons), inventing games, and creating new and more efficient ways of doing things.

- Very little are Indigos interested in aggressive sports such as football and hockey. They would rather spend their physical exercise time and energy in personal achievement and outdoor sports such as track and field, boarding, mountain climbing, cycling, kayaking. They are also attracted to discipline and self-defense sports such as fencing and martial arts.

- Because they feel so foreign to this planet, a very high percentage of Indigos has been put on anti-depressants to make them appear "Normal" and fit into our society. This is just a temporary fix, though and only adds to their challenges.

- Feel like they could be the characters on the 1980's series, "The Misfits of Science," or one of the young people in Xavier's school for the gifted in the recent movies from, "The X-Men," comic books.

- Many Indigos are drawn to theatrics, drama, and stand-up comedy. In these venues they can "pretend to be someone else," when actually they use this as an outlet to vent and express their own views and pent-up emotions. It is also a place for "misfits" to find a place of refuge and fit in.

- Because they feel so "alien" here, many go through periods of severe grief, loneliness, and displacement…and may turn to drugs, alcohol, attempt suicide for a way out.

One trademark that a high percentage of First Wave Indigos have is living through extreme hardships as children, teenagers, and young adults. Many were born into family situations that were physically, emotionally, spiritually, and psychically abusive. These Indigos have to figure out how to balance and keep their inherent integrity levels while being subjected to painful and life-shattering experiences.

A large percentage was implanted in such horrendous situations as: organized crime, physical abuse, sexual abuse, and even ritual/cult abuse and mind control. It is also common for First Wave Indigos to have some kind of alien encounters. Many have been the recipients of a "Shove-In" because of their empathic abilities. This can add to the pain and insanity in their lives. (See Laura Lee's book, "Kryahgenetics," for more information on Walk-Ins and Shove-Ins)

Other DNA Evolution

Along with the First Wave Indigos, a new mutation of our DNA is occurring. Startling evidence of a new evolution of humans has been revealed and how the end of disease may be at hand. It appears there are two other kinds of children emerging in the world today that has been identified as the "Super Psychic Children of China" and the "Children of AIDS."

The first one, the "Super Psychic Children of China," was discovered in 1984 when a child was found who was exceptionally psychic. Researchers conducted every psychic test available, and he was 100% correct, every time. Cards were turned over in another room, and that did not matter. He knew what was on every card.

Omni Magazine went to China and wrote an article on this discovery; they found one child after another who had this ability. At first, the magazine was very skeptical, that it may be a hoax, so they performed experiments. 100 children were placed in a room, and pages were removed from a book one at a time, crumbled up, and then placed under the arm of the experimenter. Each child read every word on the page. Test after test was performed, and the result was perfect. What is remarkable is that China is not the only location where these psychic children are situated. This is a worldwide phenomenon. Parents of these children are appearing and asking what to do with a child who knows everything.

Reports ten years ago in the United States revealed a baby born with AIDS. Tests a year later revealed he still tested positive. He was not tested again for six years, and to the amazement of doctors, he did not contain the AIDS virus! The staff at UCLA Medical Center performed tests and discovered that he did not have normal, human DNA. Human DNA contains 4 nucleic acids that combine in sets of 3, producing 64 different patterns called codons. The average, normal human has DNA with 20 of these codons turned on. The remainder are turned off except for 3, which are like stop/start codes on a computer. Scientists believed that the ones that were turned off were old programs from our past.

The boy in question had 24 codons turned on, 4 more than any other human being. The boy was tested to see how strong his immune system was. A very lethal dose of AIDS was placed into a Petri dish and mixed with his cells. The boy's cells remained completely unaffected. The lethal dosage was gradually increased to 3,000 times the normal infection mode, but his cells stayed completely disease free. Upon completion of this testing, the boy's blood was interacted with other well-known diseases such as cancer; however, nothing could penetrate his immune system! What is amazing is that UCLA has been focusing attention on the worldwide testing of people and have discovered that 1% of the population (60 million) have mutated to another level of DNA function!

Most of the discoveries are children; however, adults are now being tested and found to have this new, genetic enhancement. This phenomenon started ten years ago and is now spreading rapidly. I believe that we as species are beginning to mutate to another level in our history. Just as other species of life change to survive their environment, we are following suit, preserving our inhabitance. It appears that this momentum of immunity has locked itself into the grid and is being activated into the collective consciousness of humanity. I believe that this natural evolution of our DNA will be our medical savior and end disease as we know it.

I am not surprised that this is happening. Our natural evolution is going to bring great change, even if it is gradual. Within the last sixty years, more advancement has been made in all areas of human endeavor than in the history of humankind. We are discovering more about the mind, the brain, our emotions, and human consciousness in total. We as humans are beginning to understand ourselves with greater perception, along with our perceptions of our environment. The worldwide human rights movement has made a major impact on the manner in which we perceive each other. We are becoming less of separatists, and we are learning and applying our new discoveries on a moment-to-moment basis.

When we take the responsibility, individually, to send out a wave of kind action into every human interaction, then we will be able to heal the world from our inner spirit. No longer will we be entangled in the web of the polarity spider. The feeling of unity will displace the archaic brain conditionings of hate and negative emotions. By the act of these beliefs, our mental and physical systems will heal themselves as a result of a new frequency vibrating from our spirit. Again, remember that "Spiritual DNA" is our vibrating origin. This is the challenge that we all must meet. We possess the potential, individually and collectively, to direct our momentum and fulfill our human destiny of interconnectedness.

CHAPTER 11

FABRIC OF THE UNIVERSE

> There is no royal road to anything.
> One thing at a time, all things in succession.
> That which grows fast withers as rapidly:
> That which grows slowly, endures.
>
> -   J. G. Holland

Refuting Bending of the Fabric

As you have observed previously, the frequency of Spiritual DNA extended Itself spatially to create what we know as the universe. Most scientists refute the existence of God, therefore, the possibility to utilize this factor in establishing a solution to the mystery of the universe is eliminated from their equation.

Fortunately, my depth perception allows me to create a scenario that brings a sensible alternative to the table, one that is the only valid conclusion that relates to existing scientific experimentation.

The fabric of the universe is of the lightest density of any of the physical universe's dimensions. It is an extension of Spiritual DNA in the form of frequency and the closest to God's spiritual dimension. This frequency cannot be accessed by denser dimensions. It cannot be bent by forces beyond it in greater density dimensions. In other words, any of the three physical world dimensions (Sub-atomic, atomic, 3rd), and that includes revolving planets. A dimension of greater density cannot enter a dimension of lesser density; however, a dimension of lesser density can infiltrate a dimension of greater density, due to the size of the particles that make up that dimension. God's pure light frequency vibrates into all spatial dimensions by first creating the fabric of space and a universal grid through which It transfers Its governing laws through frequency waves and thus establishing Its Spiritual DNA throughout all physical world forms.

The following diagram illustrates a scientific belief that the Earth bends the fabric of the universe. Again, the fabric of space, consisting of the lightest, universal density, cannot be bent by the Earth's third dimensional density; even the Earth's smallest sub-atomic particles, for instance, if they would happen to be of the same density as the fabric, they would only pass through the fabric's particle makeup but not bend it.

76.

Einstein's Relativity Theory of Bending Fabric

Einstein's proposal is that the fabric of the universe is bent as the Earth's spin pushes against and displaces it. The fabric of the universe is the lightest of the density dimensions, even lighter than the sub-atomic particles. It is bordering

God's spiritual dimension of Pure Light. Page 13.

From the diagram, the Earth appears to bend the fabric, Earth being displayed as a dimension of the same density as the fabric; however, the fabric is of a much lighter density than the third dimension of Earth and the Earth's inner dimensions of atomic and sub-atomic particles and also contains the power of gravity. Therefore, the density of Earth's mass and velocity cannot affect the fabric's density; it only passes through it.

The Fabric Contains Gravity

It is gravity that is the same density as the fabric of the universe. Gravity originates at the same position as the fabric – at the point of Spiritual-Matter Transference in order for its control to be universal. It is gravity that gives uniformity and non-randomness to the structure of all physical world creation. It is logical that this is so. When Spiritual DNA extended Itself spatially, there had to be a force created that would allow the physical matter, later created by the spherical waves in continuous space, to be held in check.

Gravity allows our solar system to move with a sequence of rotations and revolving, establishing paths of momentum for physical objects such as planets, stars, suns, and moons. In the deepest dimensions of all physical world matter, gravity holds its position at the very core. When Earth travels its elliptical path around the sun, it is gravity from within the fabric that stabilizes all that is beyond it. It is gravity that maintains the masses of the universe's three dimensions.

CHAPTER 12

TRUST YOUR DESTINY

It is not in the stars to hold our destiny but in ourselves.

- Shakespeare

Destiny

What is destiny? As defined in Webster's dictionary, destiny is a fixed order of things established as by a Divine Decree or by an indissoluble connection of causes and effects. Indissoluble connection means something that cannot be dissolved and linking itself to the eternal series of never-ending events of the physical universe. Every particle of matter in the universe has a destiny. As every moment unfolds, it becomes an effect and at the same moment establishes itself as the cause for the next moment ("Bi-Mass/ Velocity Factor"). Although these events appear to be taking place randomly, there is an order to this grand action.

You have heard the saying, "Everything happens for a reason." More times than not, you do not realize this truth the moment it occurs. When something happens in your life that you approve, you are engaged in happiness and do not stop to think of it happening for a reason, but when something occurs in your life that you disapprove, you are engaged in unhappiness, expressing one, negative emotion after another. The next time something happens that makes you unhappy, no matter how upset it makes you feel, stop and say to yourself, "All right, I do not approve of this, but I am going to let it go and see what happens. I am going to trust my destiny." When you decide to take this path of action, you are in essence yielding to Divine Decree and acknowledging that you are just a minute particle experiencing the eternal series of never-ending events of the physical universe.

God wants you to be happy; all frequency innately contains the intent of a love vibration. It is up to you to realize that, in order to be happy in this world, you have to place God on your side. And the way to accomplish this is to accept everything that you experience. I realize that it is practically impossible for our finite minds to totally absorb this concept of acceptance 100% of the time. Each of us has our own little desires to achieve, and we are going to bring them to fruition according to our own timetable, no matter what, right? You must understand, first of all, that this type of view of how the world works is a complete aberration. This is why so many people are unhappy. This false sense of consciousness has infested the human race for millenniums.

You may have heard the phrase, "We are trying to find perfection in an imperfect world." Whoever made up that one is a complete fool. Human beings made up the word imperfect. Before humans began to attach labels to everything, there was no such thing as imperfection. Everything just was as it was, and it still is. Everything is God's perfection. We are the imperfection-minded ones.

We try to accomplish our goals within a certain framework, time. What is this concept of time? As defined, time is a period between two events. It is, however, only the perception of the person perceiving the two events. There are countless events occurring throughout the universe at the same moment, so there is actually no time between any two events.

Again, we devised the concept of time. We needed, for the sake of civilization, a system that would allow for the division of our planet's ongoing changing of day into night. Afterwards, we divided further into hours, minutes, and seconds. You must understand that in other parts of the universe where there is no sun with an earth revolving around it, there is no time, only being. There is no observer witnessing the period between two events.

As you learned in Lesson 1, there is only the action of momentum. Time does not exist, except in our mind as the observer. Therefore, to live in freedom in your mind, it is necessary to cast out the concept of time and realize that your destiny is created by the co-existence of momentum and the present. Nothing else should exist.

I know that you must call upon "past" events to make decisions in the present, but this recalling surfaces automatically from the subconscious memory bank; they are not considered the past. They are a part of the continuous momentum in the present of your destiny. The most important truth that you must remember is that you are of spirit, and for spirit there is no time. Spirit is eternal; only the flesh of our body ceases. Only the body, with its physical, mental, and emotional components, is subjected to the illusory element of time.

With each momentum that unfolds, there exists a multiple of realities occuring, simultaneously. Due to the infinite, spiritual realm of God that exists beneath all physical realms, all realities are placed into action at the same time, which then interact with the lives of all of humanity. As all of humanity moves through space/time, all possible realities are constantly unfolding before it, giving everyone the chance to choose his and her own destiny.

An everyday example of this is as follows. You may be walking through a shopping mall or supermarket and before you exists numerous items in the store and action from other people all around you. You are experiencing many, possible realities, simultaneously, all unfolding before your five senses. As your momentum is directed toward a specific, unfolding reality, as a result of the choice calculated within you mind, your ongoing destiny has been established. As you collide ("Bi/Mass Velocity Theory" in Lesson 1), with the object of your choice, momentums are swapped, and the object with the lesser engagement proficiency is directed elsewhere. In this case it may be another person that you were attracted to. You both met, exchanged momentums through speech, gestures, etc. The person with the lesser of these engagement actions will eventually submit to the other person and discontinue his or her action.

The point of this is that you must be responsible for the decision you made to direct your momentum to a potential reality of your choice. Even though the decision may have not unfolded in the manner in which you approve, it was based upon everything that had occurred before with your previous momentum choices. In any event, all choices, whether acceptable or unaccep-- table, are with aim or purpose and necessary in the continuous unfoldment of universal events.

As the universe unfolds, our consciousness evolves, and we embrace the principle of acceptance. We as a human race will gradually diminish the infliction of misery upon ourselves and heighten the degree of happiness that we constantly seek. Every particle in the universe has a destiny. Every particle of our being has its own destiny, which comprises every action in which we participate. It is the individual who realizes this truth who will ultimately gain access to true freedom in this physical world.

Learning to live with God on your team enhances the degree of positive interaction that you experience with everything around you. With God on your team, the entire force of God's presence permeates all that you encounter and rewards you with endless, little strokes of good luck. Even when something appears unfortunate at the moment, your acceptance of its appearance will

Divinely reverse the illusion that you created and reveal its true meaning. When you continuously engage in this behavior and experience the results, you will choose to live in this God Consciousness, never to return to your former, illusory perception of life.

CHAPTER 13

REVERSE GRAVITY VISUALIZATION

> Not in the clamor of the crowded streets,
> Nor in the shouts and plaudits of the throng,
> But in ourselves are triumph and defeat.
>
> - Longfellow

In 1980, I was experimenting with my new concept that I hoped would enhance the learning process in both sports and education. I discovered, after three years of experimentation, that the system of "Reverse Gravity Visualization" increases and accelerates the learning process. I decided to use this discovery as the dissertation for my doctorate of metaphysics. In 2012, I contacted every state school superintendent and proposed this system, to enhance learning capabilities. Every superintendent replied with the same message, "The state does not mandate what the local school districts teach." Plan B: I contacted all the local school districts. They replied with, "We do not accept teaching systems outside the school curriculum." I replied to all, "We should not be surprised that the United States ranks 25th in math and 17th in science." In 2017, I proposed the system to Betsy DeVos, Secretary of Education. No response.

Conventional visualization techniques usually involve the client relaxing in a recliner chair or seated in a lotus position. Because visualization is of a subjective nature, each individual learns according to his or her own capabilities. Methods of inducing visualization create the pathway for client success; however, these methods have their limitations with regard to accelerated learning. To understand these limitations, it is necessary for me to explain how the "Reverse Gravity Visualization" model functions. To accomplish this, the subject of gravity must first be discussed.

Gravity is an invisible force that attracts two objects toward each other. As is our case here, this attraction is a downward force. This downward force prevents all objects on Earth from drifting out of control; however, it also makes some disastrous claims. On the human body for instance, gravity tugs and pulls at our systems, constantly. This causes the entire makeup of our physical structure to slowly and continuously misalign from its original position. The joints compress toward one another and eventually create disease and immobility. The spinal column is pulled downward, causing the vertebrae to compress, thus causing the origins of the nerves attached to the spine to malfunction. This, in turn, affects the muscles and organs located at the ends or insertions of these nerves and cause them to operate at a lesser capacity.

When the skeletal structure begins to collapse after so many years, a similar fate occurs with the muscles attached to the bones by their tendons, along with ligaments that hold bones together with adjoining bones. This cause and effect process eventually will degenerate one's entire being: physical, mental, emotional, and spiritual. Autopsies performed on aged people have revealed that there exists a gap between the brain and skull cap created by gravity's constant pull. These are just a few of the ways in which gravity plays the role of culprit.

When the spinal column adjusts itself, there occurs an enhancement of motor neuron (nerve cells; structural units of the nervous system) efficiency between the brain and the nervous system of the spine. In addition, there is an increase of oxygenated blood flow to the brain, enhancing the functions of the brain cells and thus opening the subconscious mind hologram with greater acceleration. Results: accelerated learning.

"Reverse Gravity Visualization" counteracts gravity's destructive ways. By practicing the reverse gravity position, one can completely reverse this action by allowing gravity to work in a positive manner with the body by repositioning out of aligned systems back to their original arrangements. It is at this juncture that the difference between conventional visualization techniques and "Reverse Gravity Visualization" occurs.

If you have ever experienced any symptoms of heart attack, back injury, high blood pressure, or eye problems, please consult a physician before attempting this procedure. I do not propose that a beginner adopt high levels of declination at first. Lower levels of adjustment need to be experienced before graduating to advanced positions. This will allow one to become accustomed to increased blood flow to the brain. It also permits the proper adjustment of breathing to take place and the loss of fear due to experiencing the reverse body position. In turn, this creates increased states of relaxation, thus inducing deeper states of "Reverse Gravity Visualization."

Varying degrees of "Reverse Gravity Visualization" can be achieved with the use of a sit-up board containing numerous levels of adjustment. The more levels the better. Four levels or rungs are ideal before advancing to the 90 degree, vertical position. I suggest that one not use more than five levels; it is extremely difficult to mount a board at an angle of more than 56 degrees.

The sit-up board's foot attachment may feel somewhat uncomfortable at first; however, with practice it will feel almost non-existent. Once you have mounted, take a moment to allow your body to adjust to the downward pull supported by your ankles. Proceed with the proper breathing technique you learned in this book, and picture your images within the "Sixth Psychic Center of Consciousness" (Forehead). As you learn to relax more and feel comfortable in this position, you will consciously be aware of the areas of your body releasing their tension and resistance. Begin with the ankles. Then, proceed upward to the knees, thighs, hips, waist, chest, shoulders, arms, neck, and face. Picture the tension as a ball flowing upward through your body and out the top of your head and disappearing. You may now begin your visualizations.

As you advance to the extreme levels, you may want to experience the 90 degree, vertical position. This can be accomplished with the use of "Inversion Boots" and "Inversion Bed" made popular by Dr. Robert Martin of Pasadena, California. Dr. Martin created anti-gravity boots and inversion beds to aid his orthopedic patients in their quest to overcome back injuries.

CHAPTER 14

MOLECULAR BAND-AIDS 77.

There is no ghost so difficult to lay away as the ghost of injury.

- Alexander Smith

As my mother's best friend lie dying of breast cancer, I, as just a child, wondered of the possibility of just placing a bandage on the cancer and wait for it to heal. After her death, my image of that event returned often to my conscious mind. As I grew into adulthood and furthered my education concerning the body and its functions, I realized that healing was more possible than I was lead to believe. I began asking questions as to why no one could find a cure for cancer and other diseases. In 1992, I created a blueprint for healing on a molecular level.

Disease begins growing in an area of the body where healthy tissue exists. At what point in time did the disease actually take form? What contributed to its growth? Are there ways to prevent unhealthy forms from materializing in the body? Could the infected area be transported back in time when it remembered no disease?

In April, 1990, the Hubble Telescope went into orbit, its mission, to take photos of the cosmos 14 billion light years away. After glitches in its lenses prevented it from accomplishing its goal, the Hubble had to be placed on hold.

New discoveries from outer space? Unlocking the mysteries of the universe? Metaphysicians have known for thousands of years what the Hubble will reveal to science and medicine, that the macrocosm and the microcosm function with the same underlying, universal axioms.

All truth, knowledge, and being are interrelated, whether it is found 14 billion light years away or to the -n degree into the microcosm. One, universal vibration extends throughout everything. Medicine and science, as we have experienced thus far, treats illness by dealing with the effects of the disease instead of the cause. They see the picture only from the outside, not from within. Treat the effect and the disease continues. Treat the cause; kill the disease at its source.

In contrast to other areas of science and medicine, the field of molecular biology is in the infant stage. In 1975, only 500 molecular biologists searched for healing's secret answers. Today the figure has reached 25,000. Biological studies involving the molecular level function according to certain laws. It is now understood by certain scientists that the laws of the mind and emotional processes are correlated to the quantum field (Operation and distribution of particles at the atomic and molecular levels).

Every cell in the body has its own memory. Cells create other cells and pass on their DNA. Cells have recall to moments before a disease overcame its environment. A cell's whole being (Arrangement of molecules) once had an injury-free atmosphere until the correlating processes of the mind and emotions introduced vibrations (Wave Structure of Matter) of a negative nature, which were foreign to the cell's own microcosm.

Thoughts are things; they have their own vibration which extends outward into the atmosphere as well as projected inward to our inner being, even to our atomic and molecular divisions. Thoughts are things that originate in the mind. Via motor neurons, thought messages are transported to specific parts of the anatomy by way of intricate nerve freeways that make Los Angeles look like a rural route.

Human beings are psycho-physical units, meaning the mind controls the body, and the body controls the mind. In order for healing to occur, the two must be integrated. Integration is accomplished through rhythmic breathing, which creates relaxation of the body and calmness of the thought processes in the mind. This simple behavior modification technique is the foundation to the three-part healing equation.

Let us begin by realizing stage one, molecular nutrition. Molecular nutrition? Within every cell of the body is an engine called the mitochondria. A cell's functioning power depends upon its engine's condition. Just as an automobile's engine needs clean oil and energizing by means of a tune up, so a cell demands similar care. The cell's demands are satisfied by enzymes. Enzymes are bio-energetics, nutrient units bound together by electric life energy. The blueprints and mechanisms for creating enzymes are contained in the life plan of living cells. Science has not been able to duplicate this intricate process, so it is often reluctant to acknowledge the creative life-force of enzyme-live foods.

Every cell is made up of enzymes. Without enzymes we do not exist. Enzymes are produced only by life processes, and each part of the body requires different enzyme. For instance, the heart requires co-enzyme Q10 for its creation and release of the life energy force. The cells and the tissue which the cells comprise in the heart are Q10 enzyme. Another enzyme is SOD, super-oxide dismutase. It is the power of SOD that enters each cell and detoxifies the poisons that have accumulated as a result of improper food intake (Highly processed food and other pollutants).

The liver is the main manufacturing plant of enzymes. When the body's ability to manufacture enzymes is reduced due to aging or abuse, the immune system weakens and is made susceptible to disease. When the liver and other enzyme-producing areas of the body are cleansed and kept pure, enzymes of greater quantity and quality are made available. As a result, the cells of tissues and the tissues of organs become better qualified to perform their work. The over-all affects - prolonged life expectancy of bodily functions.

The second part of the three-part equation concerns the human, mental headset. We will be utilizing a division of psychology called Transactional Analysis to focus upon aberrations in personality. Also, "Holographic Psychology," will be utilized in this procedure.

Transactional Analysis pioneer and friend from my Purdue University days, Dr. Taibi Kahler, says that T.A. locates negative aspects of personality developed during childhood and adolescence, even sensory programming received in the fetus. As I have mentioned previously, sensory vibrations of these hazardous encounters are registered in the cellular structure of human beings and filed in the subconscious mind hologram vault, where they act as mental tape recordings for future use. Recordings detrimental to well-being, when repeated continuously over time, cause a disease vibration that generates into the future of a person's life, creating a death scenario within the psycho-physical unit.

In conjunction with Transactional Analysis, a specific, mind technique called Age Regression will penetrate the past of an individual and enable one to deprogram then reprogram mental tape recordings. During Age Regression, the memory contained in the affected cells is contacted by means of a series of visual imagery and scientific, healing affirmations. This entire process is accelerated through computerization. At this point, the laws of mind and emotional processes correlated to the quantum field and governing the operation and distribution of the particles at the molecular level are put into full force.

Holographic Psychology is very functional in this process. By evolving an individual through three levels of reality perception, a shift in the biological system will occur. This shift takes place as a result of archaic brain conditionings being displaced with new reality. Accumulation of old, destructive, mental/emotional, vibration deposits in the cells are replaced as a result of new, healthy, mental/emotional vibration created by lighter wave frequency.

A neural program is the main theme here, written to stimulate the activity within the cells. The quality of censoring is of the utmost importance at this phase, which relates back to stage two, focusing on Transactional Analysis and Holographic Psychology. Utilization of the latest neural computer technology contributes the lion's share of the healing process during this final phase, interrelating the capabilities of the neural computer with the laws of mind and emotional processes as correlated to the quantum field, bringing about an augmentation of healing in the molecular structure of the cell.

In order to realize how this delicate and most difficult procedure transpires, we must look inside the human brain. Thoughts within the brain on a mental plane manifest into subtle waves, each having a different level of frequency. A brief description of each of the four wave patterns is in order at this time. The first and most frequently used brain waves are the beta waves. During normal, wake-state hours, conscious decisions are made by betas. The brain rhythm for beta waves is an average of 21 cycles per second.

The second level of brain waves is the alpha range. Alpha waves vibrate during inner consciousness, which includes daydreams, reverie, dreams, creative consciousness, decision-making, and problem-solving. Alpha rhythms generate between 7-14 cycles per second.

Theta waves constitute the third level of wave patterns, with much more subtle vibrations occurring between 4-7 cycles per second. During theta, spiritual consciousness is the focal point. When one meditates and enters the mind, self-reflection is achieved. Reflection clears the mind of physical world residue, thus preventing further blocking of creative consciousness and allowing advancement to higher Self. Theta is also responsible for all psychic phenomena.

This brings us to the deepest level of brain rhythm, delta. The cycles of delta register no higher than 4 per second. It is the delta brain rhythm that we will use in conjunction with neural programming, to create healing on the molecular level. There exists within delta a consciousness linked to releasing all passions and desires of the physical world. At this level the powers of miracle healing unveil themselves. Miracle healing derives its function from causal conditions developed by a person's outer consciousness. Being a victim of the senses due to association and repetition, the brain becomes conditioned with limited perceptions of reality, thus affecting biological and anatomical conditions that augment over time. These conditions worsen, as conscious scenarios repeat themselves in the form of negative emotions and stress, as a result of our high-paced society.

High frequency beta waves operating according to the laws of mind and emotional processes are correlated to the quantum field and vibrate into the memory of the cellular structure, creating chaos. Once disrupted, the natural alignment of the cell begins a disintegration phase that continues until a disease mode exists and then continues to spread over a period of years (Average 20-25 years). With millions of cells infected, a visible form appears in the tissue of a particular organ.

In order for healing to occur, physical world residue must be removed from each cell using accelerated, vibrating techniques. The cells must be programmed to operate in a wellness mode. The affected cells must be convinced that the disease is of no value to itself and the function of the entire organism. It is at this point when the cells begin the transition from diseased to wellness.

This concludes the three-part equation for molecular healing. It is my strong belief that a universally, applicable, healing process will be in affect within the next twenty years of this writing (1993). More advancement will be made in the science field within the next ten years than in all of human history. We are definitely unlocking the secrets of the universe in our quest for total wellness.

CHAPTER 15

TRIAD PERFORMANCE ENHANCER

> David put his hand in his bag and took thence a stone
> And slang it and smote the Philistine in his forehead,
> And he fell upon his face to the earth.
>
> -   Bible

Autonomic Nervous System

If you ever tuned in to the Winter Games of the Olympics, you will be impressed by the biathlon event. This grueling event demands a combination of strength, endurance, and finesse. The participants must engage in an exhausting, cross-country, ski race in which, at one point along the course, they must stop and display their expertise at rifle shooting and then continue the ski race. When the participants approach the rifle competition, they must be prepared to execute the event quickly and efficiently.

One may wonder how these athletes can transform from an aggressive aspect of the sport to a passive status in such a short time-frame. There is a solution. There is a system within us called the autonomic nervous system. Ironically enough, the control center in the brain for the autonomic nervous system is the hypothalamus which also originates the emotions. This is very important, because the autonomic nervous system is responsible for regulating the involuntary action of the heart, liver, kidneys, intestines, etc. It is simple now to understand why controlling your emotions is necessary for maintaining healthy organs.

There are two sub-divisions of the autonomic nervous system, the sympathetic nervous system and parasympathetic nervous system. First we have the sympathetic nervous system. This portion of the autonomic nervous system activates those actions in our bodies that are needed to carry out vigorous activities such as the cross-country ski portion of the biathlon. This phase initiates active responses such as: increased flow of blood, muscle action, increased intake of oxygen, and other body functions necessary to ready you for competition.

On the other hand, there is the parasympathetic nervous system, which counter-balances the sympathetic nervous system, causing the body to experience a relaxed state, such as that needed during preparation and participation in rifle shooting. This calming process creates a conservation of energy during your conscious state by eliminating stress that robs your body of vital strength. The ability to trigger this spontaneous response during athletic competition is a result of practicing subconscious mind hologram programming. With the use of "Visualization Mind Technique," you can activate these systems at will and enhance your level of athletic and daily performance with better perceptive judgment and superior, neuromuscular control. Application of relaxation will induce the visualization states necessary to practice activating these two systems.

Triad Performance Enhancer

This section is an added bonus for you. I originally designed this system in 1982, to increase the performance of weight room activists and professional athletes. Recently, a gym member asked me if I took any performance enhancing supplements. He remarked that I had tremendous energy and strength. When I told him that I was seventy-two years old (2017), he was overwhelmed. I told him that the secret was internal programming.

There are three areas of the body that will give you the necessary supplement you need to enhance your performance. First of all, let me firmly express that you need to be in control of your emotions. A negative mindset will kill your efforts every time. After you practice your visual imagery of the location of each of these systems, all that is needed when you begin your activity are the words, solar plexus, spleen, heart, and your mindset will automatically activate those areas for maximum performance.

The first of the magnificent trio is the solar plexus. By activating this area, all involuntary actions are in full swing. It is the most important of the three with respect to having the most control. Located behind the stomach wall and in front of the spine, the solar plexus controls the other plexuses adjacent to it. It acts as a relay station of sorts to receive incoming messages from the brain via the splanchnic nerves. Upon receiving the messages via the motor neurons, the solar plexus reacts and sends the message to the necessary body functions responsible for your designated action.

The heart is responsible for the involuntary action of the blood throughout the body. Activating the heart's action to deliver more blood nutrition to muscle groups is key in achieving maximum, muscle participation. The third system is the spleen. This organ is a powerful tool to use for expounding energy. The spleen is located on the left, front side of the abdomen, beneath the ribcage. The spleen works as a helper in aiding the heart to provide the blood with more oxygen. The spleen disintegrates red blood cells and allows the hemoglobin (Oxygen and iron protein) to flow freely throughout the body. You now have a powerful trio to aid you in your quest for high performance during athletic and daily achievements. Activate these natural pep-pills and observe the wonders. The following illustration allows you to observe the function of the subconscious mind hologram, how it relays messages to the different plexuses of the body.

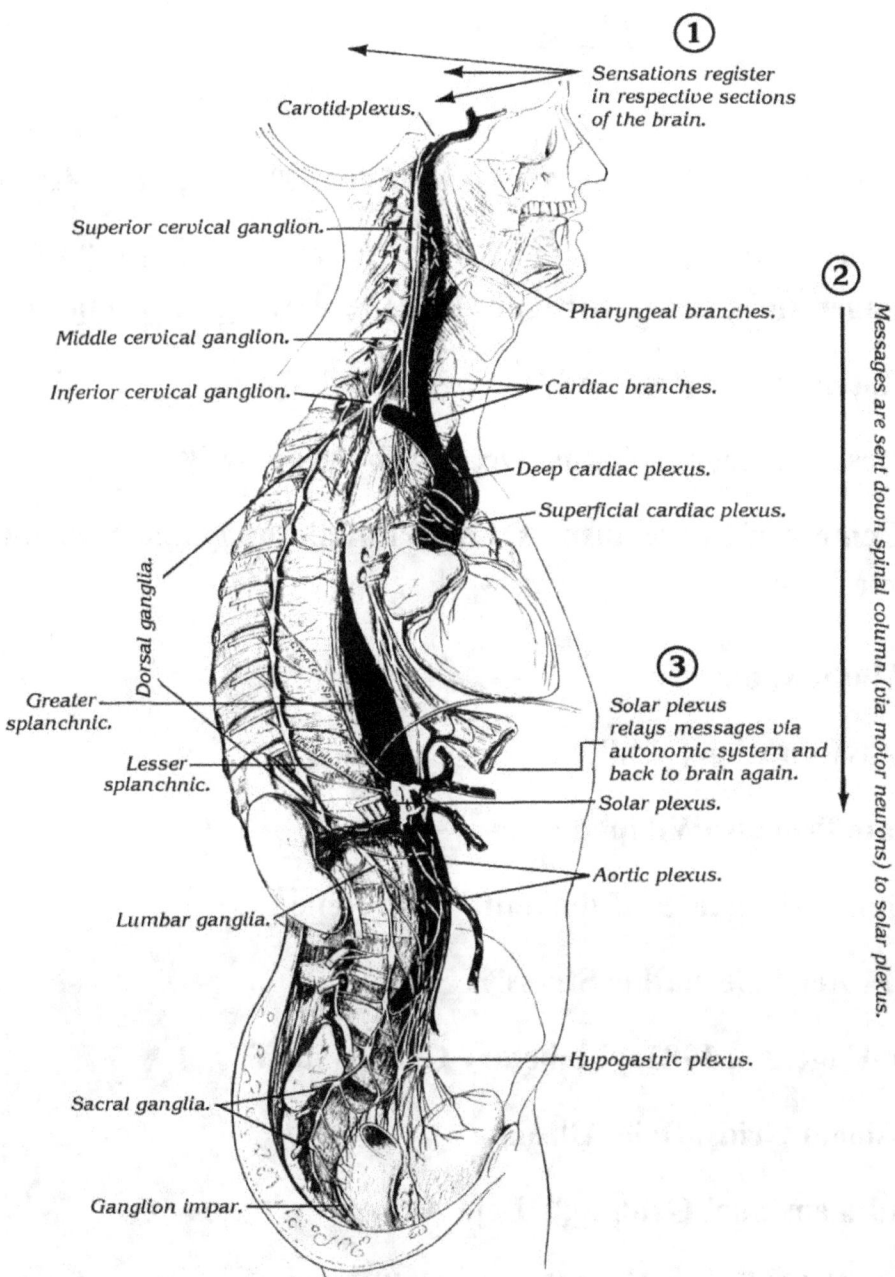

① Sensations register in respective sections of the brain.

Carotid-plexus.

Superior cervical ganglion.

Pharyngeal branches.

Middle cervical ganglion.

Inferior cervical ganglion.

Cardiac branches.

Deep cardiac plexus.

Superficial cardiac plexus.

② Messages are sent down spinal column (via motor neurons) to solar plexus.

Dorsal ganglia.

③ Solar plexus relays messages via autonomic system and back to brain again.

Greater splanchnic.

Lesser splanchnic.

Solar plexus.

Aortic plexus.

Lumbar ganglia.

Hypogastric plexus.

Sacral ganglia.

Ganglion impar.

REFERENCES

1. Elton John, "Levi, Madman Across the Water," 1971, MCA Records.

2. Geoff Haselhurst, Extension of "Wave Structure of Matter," Erwin Schrödinger and Louis DeBroglie (1927), www.spaceandtime.com

3. Geoff Haselhurst, "Time and Wave Motion," www.spaceandtime.com

4. Dr. James Pottenger, Spiritual DNA, "Holographic Psychology"

5. Seven Centers of Consciousness, Eastern Philosophy and Skeletal Anatomical Chart

6. Bible, Mathew, 6:22

7. Brain Rhythm Chart, Philip James Eastwood

8. Hologram Process, Wikipedia

9. Ascension, "Mechanics of the Shift" Lisa Renee

10. Shakura Rei, "Geopathic Stress"

11. Becker-Hagens, c 1983, "Planetary Grid System"

12. Dimensional Grids, Deja Allison

13. "Multidimensional Griding," Deja Allison

14. "Energy Grids," Gordon-Michael Scallion

15. "Ascension," Hazrat Inayat Khan, www.starstuffs.com

16. Harmonic Convergences, Dr. Berenda Fox, www.starstuffs.com

17. "The Energy Fields," Kevin Farrow

18. "Synchronicity," Wikipedia

19. "The Structure and Dynamics of the Psyche," Dr. Carl Jung

20. "The Structure and Dynamics of the Psyche," Dr. Carl Jung

21. "The Structure and Dynamics of the Psyche," Dr. Carl Jung

22. "The Structure and Dynamics of the Psyche," Dr. Carl Jung

23. "The Structure and Dynamics of the Psyche," Dr. Carl Jung

24. "The Structure and Dynamics of the Psyche," Dr. Carl Jung

25. "The Structure and Dynamics of the Psyche," Dr. Carl Jung

26. Non-Locality   Wikipedia

27. "Holographic Universe" David Bohm

28. "Holographic Universe" David Bohm

29. Non-Locality www.starsuffs.com

30. Walt Whitman

31. "Holograph Universe" Michael Talbot

32. "Stalking the Wild Pendulum"   Itzhak Bentov

33. Michio Kaku

34. "Hyperspace" Michio Kaku

35. "Simultaneous 4th Dimension"   George Bernhardt

36.   "Consciousness, Quantum Physics, and the Brain"   Quantum Mind Conference

37.   "Hyperspace"   Michio Kaku

38.   Akashic Records

39.   Buddha

40.   Arthur Koestler

41.   "The Seven, Sacred Steps"   Alana

42.   The Brain   www.starstuffs.com

43.   Silvia Cardoso

44.   "Laughter: A Scientific Investigation"   Dr. Provine

45.   "A New Concept of the Universe"   Walter Russell

46.   "Holographic Universe"   David Bohm

47.   "Holographic Universe"   David Bohm

48.   Kinesiology

49.   "Polarity Therapy Book 1"   Dr. Randolph Stone

50.   Tibetan Abbot

51.   Neville

52.   Gregg Braden

53.   Thich Nhat Hanh

54.   **Kaballah**

55.   **"The Spiritual Path"   Akhenaton**

56.   www.starstuffs.com

57.   www.starstuffs.com

58.   www.starstuffs.com

59.   **"Hyperspace"   Michio Kaku**

60.   **Innayat-Khan**

61.   **"The Book of the Hopi."   Frank Waters**

62.   **"Trimorphic Protennoia"   Nag Hammadi**

63.   **Philo of Alexandria**

64.   **"Philosophia Ultima"   Osho**

65.   **Biological Effects and Measurement of Bio-frequency/Microwaves
        E. Stanton Maxey**

66.   **Lankavatara Sutra from "Buddha Speaks,"   Anne Bancroft, 2000**

67.   **"Tao of Physics"   Chuang Tzu**

68.   **Alan Watts**

69.   **Heraclitus**

70.   **"The Wisdom of Thich Nhat Hanh"**

71.   **Innayat-Khan**

72.   **Kybalion**

73.   **Kybalion**

74.   **"Eye of Spirit" Ken Wilber**

75.   **"The Children/Adults Who Fell Through the Cracks in Our Society"
       Laura Lee Mistycah**

76.   **"Earth Bending Fabric of the Universe" Wikipedia**

77.   **Molecular Band-Aids Dr. Damon Sprock, 1992**

78.   **Anatomical Chart**

www.ingramcontent.com/pod-product-compliance
Lightning Source LLC
Chambersburg PA
CBHW081722220526
45468CB00008B/1941